DATE		

OCCUPATIONAL
DISABILITY

OCCUPATIONAL
DISABILITY

CAUSES
PREDICTION
PREVENTION

By

ROLLAND A. MARTIN, M.D.

Medical Director
Workmen's Compensation Board
State of Oregon
Salem, Oregon
Associate Clinical Professor
Department of Public Health and Preventive Medicine
University of Oregon Medical School

CHARLES C THOMAS · PUBLISHER
Springfield · Illinois · U.S.A.

Printed and Distributed Throughout the World by
CHARLES C THOMAS • PUBLISHER
BANNERSTONE HOUSE
301-327 East Lawrence Avenue, Springfield, Illinois, U.S.A.

© 1975, by CHARLES C THOMAS • PUBLISHER
ISBN 0-398-03224-6
Library of Congress Catalog Card Number: 74-18900

Printed in the United States of America
P-4

Dedicated to the Prevention of
Disabled Working Men and
Women

PREFACE

THAT OUR workmen's compensation laws are not meeting the needs of the worker is the force behind the call for reform by the National Committee on State Workmen's Compensation Laws. Our present laws came into being by each state independently creating its own brand of workmen's compensation law. At the time, each was certain it was legislating the solutions to occupational illnesses and injuries.

The solutions are yet forthcoming. The needs of the workman are not met and the revamping process wears onward in our efforts to meet the needs. Each attempt to revise the laws finds the interested parties divided over the issues; each party jealously guarding its position. As a result, our workmen find that survival under workmen's compensation insurance is a conflicting, frustrating experience.

This text speaks to the proposition that our workmen's compensation system is archaic and misconceived. The misconception began with the architects of the early-day laws. They saw the disabled workman as a person primarily needing money. Therefore, they devised a workmen's compensation system which is money oriented.

That the needs of the workman are answered by our providing more dollars continues to permeate our thinking. Thus, our attempted perfections of these laws revolve around the dollar bill. And this constricted view of the workman's needs leaves us perplexed by the maelstrom arising out of our continual efforts to purchase disability from the disabled.

A money oriented law serves only to create or worsen a workman's disability. It neither prevents nor gives him the solutions he needs to keep afloat in our society.

Any effort to reform our workmen's compensation system must provide for a greater understanding of the causes and the meaning of disability. The intention of preventing disability must be foremost in our minds as we restructure the system. Only then will we cease to leave this unending trail of disabled working people behind.

This book presents a résumé of the state laws. It brings to attention the many deficiencies that produce the adversary setting so rampant in the workmen's compensation system.

With the current result of our efforts in mind, the text next develops thoughts as they relate to injury and disability. "A theory of disability" is perhaps a better term to use in describing the chapter presented for your consideration as a discussion of disability.

Any criticism of a system or presentation of a theory of disability is useless, of course, unless a practical solution is offered. Thus, a practical model of a disability prevention program is also presented, patterned after the Oregon plan.

The Workmen's Compensation Board in Oregon established a Disability Prevention Division in January 1970. Its sole purpose is to prevent, or reduce, the degree of disability in our workmen having work-related illnesses or injuries.

The Oregon approach is straightforward and effective. Because it is a simply structured plan, other states, or insurers, or large employers can readily implement a program based upon this model.

Disability prevention better serves the need of the occupationally ill or injured workman. In meeting these needs many benefits in turn accrue to the State and its employers. Preventing a disability obviates the necessity to rehabilitate. Minimizing the numbers of disabled gives a state and its employers a more viable work force.

Disability prevention offers an alternative to disability.

ACKNOWLEDGMENTS

THE CONCEPT of disability prevention discussed in these writings is the collective result of the efforts of a great many contributors. Some of the contributions come from writers and some come from doers. Each one recognizes the potential impact that workmen's compensation has on the life of every member of our society in this nation. Each one sees one or more of the various problems besetting workmen's compensation. All are moving to correct these problems.

To writers such as those whose writings are referred to in this book, the author is indebted. It always helps to know that others are interested or concerned enough to also study and inquire into this matter of disability, to look to the needs of the disabled person such as a workman.

The move from concept to the practical application of disability prevention is the result of a considerable effort by a good many people. The Workmen's Compensation Board in the State of Oregon is a progressive one. It is composed of Chairman M. Keith Wilson, Commissioner George A. Moore, and Commissioner Gordon Sloan. Their solid backing of the concept aided in bringing disability prevention into being in Oregon.

A special acknowledgment is owed to the members of the Industrial Accident Advisory Committee who spearheaded the disability prevention plan. They are Edward Whalen, Karl Frederick, Robert McCallister, Charles Gill, Chet Diehl, William Waste, and James Smart; and, to Dr. Morris Crothers and Representative Keith Skelton.

Constantly encouraging the author to pursue the subject of disability prevention has been a good friend, R. J. Chance, Director of the Workmen's Compensation Board.

My secretary, Mrs. Maureen Amlong, has worked faithfully with me through many months. To her goes a great deal of my gratitude for her gentle but constructive criticism of this work.

R. A. M.

CONTENTS

OCCUPATIONAL DISABILITY

Chapter One

A HALF CENTURY OF EFFORT

A MOVE FROM COMMON LAW

PRESENT-DAY WORKMEN'S COMPENSATION LAW is a device of social consciousness. In theory, its sole purpose is to protect working men and women from personal disaster arising out of occupational illnesses and injuries. The beginnings stem from conditions in Europe during the 1880's. Our own laws in the United States did not become a reality until 1908 with the passage of the Federal Employee's Compensation Act.

New York State later enacted a workmen's compensation law in 1910 only to see it declared unconstitutional by the New York Court of Appeals. The Court's position held that the compulsory coverage requirement of New York's law created an imposition of liability without fault and the taking of property without due process of law. As a result of this decision, the laws later enacted became elective laws excluding certain industries and occupations, and embraced the "no-fault" concept.

Prior to the no-fault approach in workmen's compensation law, common law prevailed in all recovery suits filed by occupationally ill or injured workmen. Under common law, the employer has three defenses:

1. *Contributory negligence* which requires the workman to prove he is in no way at fault for the illness or injury;
2. *The fellow-servant doctrine* which holds that if a fellow workman's negligence results in an injury to the claimant, the claimant is not entitled to recovery from the employer;
3. *Assumption of risks* which holds that if the hazards of the job are known to the workman, there is no recovery.

The burden of proof rests squarely upon the workman when

[3]

the rules of common law prevail. The workman must prove the employer's liability.

Many workmen found fellow workmen reluctant to testify against their employers. The courts were often unsympathetic toward workmen's complaints and awards were inconsistent and often inadequate.

As a result, legal costs mounted and the wastefulness of such a system became obvious. Many workmen could not successfully carry the burden of proof through the courts. They became the burden of charity. It became a time for change through social legislation.

With this, the states gradually began enacting their present workmen's compensation laws, with Mississippi becoming the last of the forty-eight states to implement its law in 1949.

Since each state legislated its own workmen's compensation law, many variances exist today from state to state. However, a common thread of philosophy exists in all the laws enacted. Common to workmen's compensation laws is the agreement that the costs of work-related diseases and injuries are the burden of the nation's work places and its employers. Also common to the present laws is the no-fault principle. Under this concept, the workman is not burdened with proving employers' liability, nor is the employer burdened with the defense of liability suits.

However, uniformity in workmen's compensation laws is not found in the coverage of workmen, the benefits provided, the safety efforts, and the appeals process.

COVERAGE

Whether or not a workman comes under the protective coverage of workmen's compensation law in a state depends upon the law of the specific state. Deficiencies in coverage exist. Presently, coverage is limited to about 83 per cent of the nation's workmen. The factors creating these deficiencies are:

1. Elective laws
2. Numerical exemptions
3. Exclusions of specific occupations
4. Occupational diseases

Elective Laws

Elective laws exist in seventeen states.[1] Under elective law, the employer may reject the compensation act. In rejecting the act under an elective law, an employer loses the three common law defenses: contributory negligence, negligence of fellow employees, and assumption of risk.

Of the seventeen states operating under elective laws, all but the state of Louisiana require employers to insure through either a private insurance company or a state fund. Although these states require the employer to provide this insurance, the claimant workman is still faced with proving employer liability in an adversary setting in the event of a dispute.

Numerical Exemptions

Numerical exemption creates another hiatus in universal coverage. Presently twenty-three states have numerical exemption clauses in their acts.

While the qualifications for numerical exemption vary from state to state, it essentially means an employer must have a specified number of employees in his employment before any form of protective insurance or security is required under the workmen's compensation law of the state.

In those states allowing numerical exemptions, the range is from two to fifteen employees. In the majority the range is between three to five employees. Two states exempt all employers having fewer than two employees, and one state exempts those employers having fewer than fifteen employees.[1]

Exclusions of Specific Occupations

Exclusion of certain hazardous and non-hazardous occupations accounts for additional deficiencies in coverage. As an example, farm workers are excluded from coverage under either an exclusion clause or a numerical exemption clause in the laws of most states. Only seventeen states provide coverage for farm workers that meets the standards of the Department of Labor.[2]

Occupational Diseases

The matter of coverage for occupational disease also creates a further deficiency in the protection of workmen. Seven states limit their coverage of occupational disease by specifying the occupational disease they cover. The number of occupational diseases covered by these states ranges from six to forty-seven.

The time limitations wherein a workman may file a claim for occupational disease also limits coverage of occupational diseases in the states. As an example: in the state of Indiana, disability must occur within two years of the last exposure.

Time limitations for filing occupational disease claims after injury vary from as short as sixty days to as long as three years after exposure. Few states provide coverage for occupational disease in the same manner as occupational injuries are covered.[1]

BENEFITS

Compensation Income Benefits

Workmen's compensation utilizes four classifications of disability for purposes of determining compensation income benefits:

1. Temporary total disability
2. Temporary partial disability
3. Permanent partial disability
4. Permanent total disability

All states' workmen's compensation laws contain a *waiting period* requirement that determines the workman's eligibility for temporary disability income benefits. The waiting period differs from state to state and ranges from two to seven days.

In addition to the waiting period requirement, all states im-

TABLE I
WAITING PERIOD

Number of Days	Number of States
2	1
3	14
5	6
6	1
7	28

pose a *retroactive period*. The retroactive period is that period of time the workman must remain disabled to be eligible for the income benefits not previously paid during the waiting period. The retroactive periods throughout the states range from five days to as long as six weeks.

As an example, Pennsylvania has a seven-day waiting period and a six-week retroactive period. A claimant workman must first be totally disabled seven days to become eligible for income benefits. If the workman remains disabled for at least six weeks, he then receives retroactive income benefits for the first seven days of his disability.[1]

Temporary total disability income benefits are also limited in the dollar amounts a workman receives by two differing methods. The first method utilizes a percentage limitation based upon the workman's average weekly wage. These percentage limitations for calculating the compensation income range from a low of 55 per cent to a high of 90 per cent of the workman's average weekly wage in different states.[1]

In addition, all states specify a maximum amount they will pay regardless of the workman's weekly wage. An example of this limitation on income benefits is the state of Nevada. Although Nevada pays 90 per cent of the workmen's average weekly wage, this is nullified by the use of a maximum benefit clause limiting maximum compensation income benefits to $67.16 per week for the workman drawing top wages.[1]

Other mechanisms also exist which serve to limit temporary total disability income benefits for the workman. Thirty-three states limit the maximum time they allow a workman to be temporarily totally disabled. These allowable time limitations range from 240 weeks to 500 weeks.[2]

Although six states do not limit the period of time a workman may be temporarily totally disabled, they accomplish the same goal. They limit temporary total disability by setting limits upon the total dollar amount paid to any workman temporarily totally disabled.

Temporary partial disability income payment is a device used wherein the workman may resume working for his employer at less than his average weekly wage and also qualify for a percentage of the temporary total disability income payment to compensate him for his temporary partial wage loss.

Permanent partial or permanent total disability income benefits are those which continue beyond the termination of temporary total or temporary partial disability. It is possible, however, to have a situation of permanent partial disability in a workman without any preceding time loss due to the injury.

The dollar amounts of permanent partial or permanent total disability income benefits are determined by the part or parts of the workman's body injured. Legislators long ago decided they could predetermine the worth of certain parts of the body. Thus, they devised a *schedule of injuries* assigning specific dollar values to a specific body part when injury to that part resulted in permanent damage.

While variances exist from state to state, the body parts falling into a classification of scheduled injuries in all states are the extremities, the eyes, and the ears.

Rather remarkable differences of opinion exist among the respective states regarding the specific values of scheduled injuries. For instance, Arizona pays $33,000 to the workman losing an arm at the shoulder. Idaho pays $17,851 for the same injury. However, Massachusetts pays only $6,750 to the workman.[1]

Injuries to all other parts of the body not listed in a schedule by a state are unscheduled injuries. Here we see a large variety of injuries such as back injuries, central nervous system injuries, and internal injuries.

Specific dollar values are not pre-assigned to unscheduled injuries resulting in permanent disability. In addition to the permanent impairment of the injured part, age, education, and sometimes other factors, enter into the determination of the workman's degree of permanent disability for unscheduled injuries.

Permanent partial disability income benefits are determined by the dollar amount of the permanent partial disability award. This income benefit is commonly paid to the workman over a period of months or years, depending upon the total amount of the award. If the award is small, the workman may receive the award in one lump payment.

Permanent total disability awards, like permanent partial disability awards, result from either scheduled or unscheduled injuries. Permanent total disability in the workmen's compensation acts is commonly described as found in Oregon's Workmen's Compensation Law 656.206:

Permanent total disability. (1) As used in this section:

(a) "Permanent total disability" means the loss, including preexisting disability, of both feet or hands, or one foot and one hand, total loss of eyesight or such paralysis or other condition permanently incapacitating the workman from regularly performing any work at a gainful and suitable occupation.

Permanent total disability income benefits are not necessarily paid for the lifetime of the workman. Although the workman may be obviously totally disabled for life, fifteen states do not pay for life. They either limit the length of permanent total disability allowed by time limitations, or they limit the total dollar amount available to the workman. Some states do both. In all, there are fifteen states that limit permanent total disability benefits for a workman to less than ten years.

As with temporary total disability, the number of dollars paid to a workman weekly as permanent total disability income is limited in all fifty states. Here again, a percentage formula of the workman's weekly wage is used, and this is further limited by the use of the maximum dollar amount allowable in any instance.

Maximum weekly benefits vary widely. Arizona pays as high as $150 weekly. Alabama pays $60 per week. The lowest three are Louisiana, Mississippi, and Texas, where the maximum is $49 per week.[1]

Medical Benefits

A major benefit provision in workmen's compensation law is medical care for the occupationally ill or injured workman. The intent is to provide prompt medical care of the highest quality.

Past history of workmen's compensation contains evidence that the workman did not always receive the best quality of medical care, nor with any degree of promptness. Company owned and operated medical departments often left much to be desired in past times.

The gradual trend resulting from this questionable medical care of the past is the *free choice of physician* concept in states' workmen's compensation acts.

Proponents of the workman's right to choose his or her physician maintain that the workman receives better medical care with greater promptness, and that the physician is not "pro-employer" in dealing with the workman. Although arguments exist for both sides, the trend is still toward a free choice of physician. At this time, twenty-five states allow the injured workman a free choice of physician.[2]

In spite of many changes in medical benefits, limitations of benefits still exist in some workmen's compensation acts. This occurs mostly in the length of time medical care is provided and in the total dollar amount spent. Such arbitrary limitations of medical benefits still exist in eleven states.

In addition, other limitations exist in certain other aspects of medical care provided in seventeen states. Some do not provide payment for care in rehabilitation centers or in home health care plans. Some refuse payment for care provided by occupational therapists, osteopathic physicians, registered nurses, or psychologists, according to the National Commission on State Workmen's Compensation Laws.

Rehabilitation Benefits

Physical and vocational rehabilitation of occupationally ill or injured workmen is provided with varying limitations by the various states. Thirty-seven states have the provisions for rehabilitation written into their workmen's compensation acts. Four states operate their own physical rehabilitation centers.

Twenty of the thirty-seven states reporting their provisions for vocational rehabilitation have limitations upon the length of training or the total dollar amounts provided for vocational training.

A major problem in rehabilitating a workman is the workman's need for financial assistance during the period of rehabilitation. Currently, special maintenance benefits are provided in only twenty-seven states.[2]

Secondary Injury Benefits

Second injury funds serve to benefit both the workman and the employer. The purpose of second injury funds is to:

1. Encourage the hiring of the handicapped;
2. Protect the employer by allocating costs of a subsequent injury which, when combined with a pre-existing handicap, causes a greater disability than would have occurred had not the handicap pre-existed.

All but four states reportedly have some form of a second injury fund. Various and sundry limitations exist in the laws on second injury funds from state to state.[1]

One-half of the states limit the use of second injury funds by describing the type of injury a handicapped workman must suffer before the employer is eligible for second injury relief. For example, Alabama limits second injury fund relief for the employer to those injuries involving the loss of an eye, leg, arm, foot, or hand.[1]

Additionally, twenty-seven states specify the degree of disability a workman must have as a result of the combined effects of the injury and the pre-existing handicap before the employer is entitled to second injury relief.

SAFETY

Prevention of occupational illness and injury functions at varying levels of effectiveness among the states. While there is a long history of agreement that good safety practices are needed, the question of whose function it is to insure that these safety practices are followed has hampered the effort.

Many large employers operate their own safety programs because of their worth to the company. Many insurance companies provide safety advice and services to their customers in the form of inspection, engineering, and education.

Among the states, many entities are involved in accident prevention programs. It is difficult to evaluate what influence workmen's compensation has on safety practices.

The implementation of the Occupational Safety and Health Act of 1970 is a result of generally ineffectual safety practices. In those states not adopting the federal safety standards, the functions of safety and accident prevention are now being assumed by the Federal Government.

APPEALS PROVISIONS

All states have provisions for appeal within their workmen's compensation acts. In the beginning of our present-day workmen's compensation laws, the designers of the no-fault concept believed the laws would be self-administrating if they detailed the benefits of the law. Therefore, they believed such laws would markedly reduce the need for litigation.

However, all workmen's compensation systems are involved in considerable litigation today. Workmen and employers alike find they still must resort to the use of private counsel to settle their disputes in the courts.

The right to jury trial has all but been eliminated in workmen's compensation. Five states still allow jury trials. The efforts to eliminate jury trials are based upon the costs and the delays seen in bringing the case to trial under the jury system.

Appeals in forty-five states are first referred to adjudicators in the workmen's compensation system. If they fail to settle the issues, the matter is then appealed to the courts. Such legal disputes in five states are referred directly to the courts without any interceding body in the process.[1]

Time allowed for appeals varies markedly among the states. Appeal time ranges from as short as five days to as long as seventy days.[1]

Also, the court to which a dispute is appealed differs from state to state. Disputes in ten states are appealed directly to the Supreme Courts. Other states also allow appeals to the Supreme Courts, but they must first pass through the lower courts.[2]

SUMMARY

The perfect workmen's compensation law awaits its author. It is not yet written. Models of proposed ideal acts exist. In general, these models are not far apart. The difficulty lies in getting men to agree. The effort of this century has been to improve the workman's benefits and to perfect self-administering workmen's compensation law through legislation and litigation. Much of the legislation is the result of reluctant compromises between labor and industry. And, the gains are slow.

Chapter Two

THE EFFORT AND THE RESULT

THE MISCUE

IN SPITE OF OUR EFFORTS, workmen's compensation systems are filled with many faults. Outstanding among these faults is an overwhelming tendency to focus our attention upon the money factors. Money, it seems, is at the forefront of our thinking. So much so that it clouds our vision and our ability to recognize a workman as an individual having individual needs.

Just as it is with you and me, the needs of the workman are factual and highly individualistic. It is not always the dollar bill that is the greatest need.

Because the entire past of workmen's compensation is based upon a money-oriented philosophy, our minds remain mired in this concern over the costs of the system to this day. In turn, this causes us to approach any improvements in the system as though we are developing a dole system. As a result, the basic hopes underlying the creation of workmen's compensation remain mostly in the offing. Although it is more than half a century since we implemented these first laws, workmen still suffer personal disaster from occupational illnesses and injuries.

The objectives needed to overcome the present faults in the system are already set down. Business, labor, governmental bodies, and many associations have published objectives they believe necessary to create an equitable workmen's compensation system.

Interestingly enough, the objectives stated by the different parties are all strikingly similar. Those published by the U.S. Chamber of Commerce pretty well embrace these recommended objectives.[1] The objectives put forth by the U.S. Chamber of Commerce are six in number and are designed to:

1. Provide sure, prompt and reasonable income and medical

benefits to work-accident victims, or income benefits to their dependents, regardless of fault;

2. Provide a single remedy and reduce court delays, costs and work loads arising out of personal-injury litigation;
3. Relieve public and private charities of financial drains incident to uncompensated industrial accidents;
4. Eliminate payment of fees to lawyers and witnesses as well as time-consuming trials and appeals;
5. Encourage maximum employer interest in safety and rehabilitation through appropriate experience-rating mechanism; and
6. Promote frank study of causes of accidents (rather than concealment of fault), thereby reducing preventable accidents and human suffering.

These objectives fulfill the basic intent of workmen's compensation law. Essentially, all parties are in agreement with these objectives. But, when it comes down to implementing these objectives, we see the miscues. General disagreements exist in each of the fifty states.

LIKE IT IS

Let's examine the various segments of our existing workmen's compensation laws to see where we fail in meeting those objectives. By taking a closer look at the facts, we shall see the present deficiencies in our state's laws. Also, we will see readily enough why it is that the National Commission on State Workmen's Compensation Laws* is calling for reform.[2]

Un-Coverage

Un-coverage is the first matter. To be, or not to be, covered under workmen's compensation insurance as a workman is presently too dependent upon where you live, your occupation, and the size of your employer.

Thirty per cent, or more, of the employees in fifteen states are not covered by workmen's compensation insurance. An additional thirteen states insure no more than 85 per cent of their employees.

---

* To be called "The National Commission," henceforth.

This deficiency in universal coverage of all workmen results from states allowing permissive laws such as elective coverage, numerical exemptions, and exclusions of specific hazardous or nonhazardous occupations.

Elective laws serve mainly the employer. Such laws deny employees rights to protective coverage under workmen's compensation. As a result, in the event of an occupational illness or injury, the burden of proving employer liability lies with the claimant workman. The same problem also exists when numerical exemptions, and exclusion of hazardous or nonhazardous occupations are written into the workmen's compensation law.

The tragedy of any deficiency in coverage is its effect upon those needing the coverage the most. Too frequently, those excluded are the very poor. It destroys them because they are the very ones having no other resources.

The great bulk of the people we exclude by elective laws, numerical exemptions, and type of occupation are those people employed by public charities, or as domestic workers, or as farm workers. All are people likely living on the economic brink.

Why is it we almost universally deny this class of workmen protection under workmen's compensation? Many answers are given to this question. At the gut of this question and underlying our failure to afford them protection, may be an embarrassing national sociological-moral issue.

It nearly appears as though we are saying to those such as the farmworker, "Friend, it's all right if you starve, but our eating habits are dear to us."

We might ask ourselves, "Do we really deny coverage to farmworkers because we harken more to the wail and the cry of the farmer-employer bemoaning his costs of food production? Is our fear of increasing the costs of filling our own bellies where the problem really lies?" It appears so.

Two-thirds of the states presently do not cover farmworkers on the same basis as other workers. Even though farmworkers comprise a large work force in our nation, we can't seem to bring ourselves to facing the issue.

The National Commission, in making its recommendations for coverage of farmworkers, is even timid. While it believes the farmworker is entitled to coverage, the Commission recommends a creeping, two-stage approach: [2]

1. Each agriculture employer who has an annual payroll in excess of $1,000 will be required by July 1, 1973, to cover workers under workmen's compensation insurance;
2. By July 1, 1975, all farmworkers will be covered in the same manner as all other workmen.

Question: What does a thousand dollar payroll have to do with whether or not an occupational illness or injury is acceptable?

Farming is historically a hazardous work. Additionally, it is very likely that a smaller farmer has less safety training than does a larger, corporate farming operator. Compounding this danger is the fact that a small farmer is less likely to be inspected for the safety of his operation.

Excuse: Most states claim it isn't administratively feasible to cover farmworkers. They fear the administrative problems involved. The farmworker is too migrant.

Farmworkers cross many state lines and have many employers in a given year. This supposedly creates serious administrative problems in reporting injuries, medical care, rehabilitation, rating disability, and making proper audits for administering agencies.

In fact, there are only five states presently covering farmworkers in the same manner as all other workmen in their compensation laws. They have successfully taken on these administrative burdens. Meanwhile, the remaining forty-five states duck the issue.

Backsliding

Our two-headed approach to the poor in our nation is strange indeed. For too many years we've considered the bulk of farmworkers as a class of poor migrants and we've essentially ignored their plight. Of late, we've begun to recognize the predicament of farmworkers and others we consider as the poor.

We're now attempting a turn-around by ploughing millions of dollars into federally funded programs to upgrade their life style. Our plan is to lift those such as farmworkers out of their miserable life style, to educate farmworkers' families to a better life.

But what happens when this same farmworker has his leg crunched beneath a tractor wheel? What happens to this man and this family we were trying to lead out of poverty and degradation prior to his injury?

Does our other head now give the farmworker and his family a practical lesson in human dignity by denying them equality under workmen's compensation? Do we leave them any hope but charity? Are our own fears over costs of proper coverage for them this strong? From close up it certainly appears that our concern is for costs, not the needs of such workmen. How can we suggest any other reason for our failure to provide coverage to employees of public charities, domestics, and farmworkers? It is time that we stop backsliding. The cost of workmen's compensation coverage for employees of public charities, domestics, and farmworkers is not the sensible factor for us to consider. It represents faulty thinking on our part.

Denying this group of workmen protection under workmen's compensation law does not keep the costs of operating charities, or maintaining our homes, or the costs of foodstuffs at a lower level. We merely transfer those costs for their care to welfare by our present methods. We simply bury the true costs of their illnesses and injuries out of sight.

The occupationally injured domestic, or public charity employee, or farmworker does not cease to exist with injury. The costs of relegating them to a welfare status far outweigh any costs of properly protecting them under workmen's compensation. The proper coverage of all workmen is our best bet, if we really want to lower costs.

A viable, effective workmen's compensation system offers us the best opportunity for returning a workman to gainful occupation after illness or injury. This is the realistic way of lowering costs.

Un-Scene

Other problems in coverage also exist. That the following problem needs some attention may be rejected by some readers. It does need consideration, however. The problem is that of the two-job worker.

Two jobs is the life style of a great many people today. Usually, one job is the major source of income and the other a minor source of income. The standard of living is normally based upon the major income.

But what happens to the workman in the event of a job-related illness or injury while working the lesser paying job? We can use Louise as an example of this situation. She is widowed and the sole support of three children. She works full time and earns $75 a week. It isn't enough.

Louise finds her net take-home pay is less than $75 a week, so she works part time in the evenings. This part-time job brings her an extra $25 a week. It does until she hurts her back one evening.

She is unable to work either job from this point on, and Louise is told she needs surgery. She has a herniated intervertebral disc. She also finds that her compensation income check is based upon her earnings on the part-time job. Louise now has a weekly income of $40, provided by the minimum income benefit level in the workmen's compensation law of her state.

What is the solution for Louise and other workmen caught in this financial trap solely because they are injured on the lesser paying job? Do we continue to consider this as one of life's gambles? Is this attitude a sensible solution, or is it possibly our most costly answer?

Somehow we need to build a bridge of coverage to protect the workmen caught in this type of circumstance. It does little good for a workman, or a nation, to plunge him into a financial and emotional abyss because he happened to be injured on the wrong job. The bridge needed is the one giving a workman the coverage of the usual job. How this is possible, remains in the future.

Sick, Sick Coverage

Coverage of occupational disease is also incomplete. Nine states limit coverage of occupational disease to only those diseases listed in their Acts. Such listings of occupational disease cover as many as forty-seven diseases, and as few as six.[1]

The mere fact these legislative bodies limit occupational diseases to those of their choosing denies workmen proper coverage. Also compounding this matter of coverage are other limitations in seven states that deny the workman having an occupational disease full coverage for needed medical care.

BUCKS AND BENEFITS

Compensation Income Benefits

Disability income for the workman remains the great debate in workmen's compensation. This is where we find ourselves veering obliquely onto a path that consumes ridiculous amounts of our time and our thinking.

The arguments concerning income benefits range from the supposition that too great an income encourages a workman to remain disabled, to the proposition that too little income destroys the workman's initiative.

That we consider income benefits as still the greatest need of our claimant workmen is obvious by the events, past and present. It is further evidenced by the National Commission report on state workmen's compensation laws.

The National Commission devotes more space in its report to the *income maintenance objective* than to any other aspect of workmen's compensation laws. However, there is a reason for this if the mound of testimony given before the Commission is reviewed. Income benefits dominate the testimony brought before the Commission.

Dramatic examples of inequities in income benefits are given by those testifying. Senator Javits, testifying before the Commission, gives an example of a Texas workman. This man lost the use of both arms and legs as a result of a work injury. Prior to this injury this fellow earned $118 per week. Presently, his compensation income is $49 per week.

By the nature of his injury, this man is permanently and totally disabled. But the worst is yet to come for him. His income benefits will cease after eight years under the existing Texas workmen's compensation law.

The future is indeed bleak for this workman because he had the misfortune to be injured in Texas. Similar time limitations exist in nineteen other states. The decerebrate reasoning for enacting such laws is unreasonable. How is it legislators can predetermine when it is that a workman ceases to be permanently totally disabled?

Such an approach to legislating workmen's compensation law is surely with the attitude that workmen's compensation is a dole system. Their kind of shortsighted legislation creates a degrading welfare system rather than an insurance system.

Another method utilized to limit income benefits is the limitation placed upon the total number of dollars ultimately paid a workman having a permanent total disability. Currently, twenty-three states have such limitations in their laws.

What happens to the workman totally disabled in these states having arbitrary time or dollar limitations? No indications are particularly given. The number who go on welfare or turn to private charities or to some other agency for support is not known, nor is it likely to be publicized.

An additional inequity in the income benefits for workmen permanently totally disabled exists in nine states. In these states the maximum weekly income for permanent total disability is less than that allowed for temporary total disability. Imagine the reaction of a workman upon making this discovery.[1]

Temporary Total Disability

Workmen suffering a temporary total disability of short duration stand to lose two to seven days' pay because of the *waiting period* requirement for disabling injuries. This is a usual practice of workmen's compensation in all states.

The income benefits lost by a workman due to the waiting period requirement are the result of *retroactive period* requirements in workmen's compensation laws.

Dependent upon the particular workmen's compensation act, all workmen must remain temporarily totally disabled from seven to forty-nine days in order to recover the income benefits withheld during the waiting period.

A basic philosophy behind the waiting period is that it is a method of discouraging malingering. This matter of malingering is discussed in later chapters. Suffice it to say, at this point, that a waiting period never discourages a malingerer, if malingering exists.

It is in the entire area of the *waiting period* and the *retroactive period* that we see a hiatus in the philosophy of workmen's compensation law, the schizophrenia of social thought and social action.

We mouth one thing, but our actions differ markedly from our preachings. We maintain that the occupationally ill or injured person is to be compensated for his wage loss, but then, we pass laws which do not fulfill our intentions. Why? Is our fear of malingering this great? Or is it the costs of paying the honest workman his rightful benefits we fear?

The National Commission reports the seven-day waiting period, when combined with the 28-day retroactive requirement, pays benefits to approximately 83 per cent of the lost-time claims.

Reducing the waiting period and the retroactive period to three days waiting and fourteen days to qualify for retroactive benefits would cover approximately 93 per cent of the lost-time claims. This means 10 per cent more workmen having time-loss claims would be compensated in part for their lost wages due to the illness or injury.[2]

The National Commission recommends that all states shorten their waiting-retroactive periods to conform to these latter figures which are the same standards recommended by the U.S. Department of Labor. This means that a workman suffering a disabling injury must remain disabled beyond fourteen days if he is to receive compensation for the first three days of his disability.

What reasoning is it that brings the National Commission to the conclusion that the workman off work for thirteen days has less need for three days' income benefits than the workman off work for fifteen days?

The reasoning is a compromise over costs. We again find the workman's needs becoming secondary to costs by those who view many claimant workmen as nothing more than malingerers. However, the proponents of waiting-retroactive periods as a hedge against increased costs may be fooling themselves.

The true costs of the waiting-retroactive period philosophy may, in fact, be more costly to workmen's compensation than if no waiting-retroactive period requirement existed. No facts exist to confirm or deny this possibility. But, we might ask ourselves,

"Why should a workman off work ten to thirteen days resume work on the fourteenth day?"

Our thoughts throughout workmen's compensation seem bound to this same presumption that all workmen slyly wait for the *million dollar injury* and the *golden limbo of retirement*. Therefore, we must guard him from cheating us and our pocketbook.

How much compensation income is the occupationally ill or injured workman entitled to receive? Here again we find the philosophers and head-nodders of workmen's compensation law ready to renegé.

Are you aware that the law declares that it costs you less to live after having a time-loss injury? This is evident if you review the percentage of weekly wage paid to workmen as compensation income benefits. All compensation income is based upon a percentage of the weekly wage in all states.

If, however, this percentage formula gives you a greater compensation income than the legislators desire, they give you less. The law merely reduces your income benefit by setting an allowable maximum amount payable.

If we look at the maximum income benefits paid to workers in many states, it becomes unreal. For instance, how would you like the prospect of living on $49 per week? Put yourself in this fix in Arkansas, if this sum meets your living requirements. Forty-nine dollars a week is the maximum compensation Arkansas pays a workman for temporary total disability.[1]

As of January 1972, more than one-half of the states paid maximum weekly income benefits that are below the poverty level, according to the National Commission. The poverty level it refers to is the 1971 national poverty level in non-farm families of four persons earning $79.56 per week.

When we consider that the reported average weekly wage for all workers in 1972 was $150 per week, a nearly 50 per cent reduction in earnings is a critical matter to that person.

Wage loss is not the only loss to the occupationally ill or injured workman, however. Of course, this is dependent upon the length of a workman's disability. In prolonged disability, a workman stands to lose prepaid health insurance, sick leave, vacation, and his retirement benefits, in addition to being relegated to a poverty level income.

Medical Care Benefits

We injure them, we make them ill, and still we crawfish. Full medical benefits are not provided the workmen by our lawmakers.

Nine states limit the duration of medical treatment allowed per injury, or they limit the maximum number of dollars they spend per injury.[1] Of course, this raises an immediate question. What becomes of the workman who fails to meet the healing standards of legislative edict?

Certain types of medical services are also limited in some states. Some states refuse to pay for medical care rendered by home health care servitors, osteopaths, psychologists, registered nurses, occupational therapists, or rehabilitation centers.

Additionally, limitations are also placed upon subsequent care for the aggravation of an old injury. Such a limitation has the effect of turning the workman out at the time of claim closure. He no longer has any right to medical care for that injury. This places the workman in the same precarious position he is in when he signs a release agreement with an insurer after an automobile accident.

The various workmen's compensation acts may range from fair to excellent in providing medical care for workmen's physical injuries, but they run scared when it comes to matters of the mind. This results in a sorry waste of money because a workman in need of mental health care is only partially treated.

The psychological problems seen in many ill or injured workmen are a knotty subject in workmen's compensation. Because they are, the psychological aspects of injury and disability are the most ignored, unwanted medical problems existent in workmen's compensation.

The immediate concern in care of the workman's mental health is again the costs. The total costs in the workmen's compensation system for 1971 is reported at $93 billion. How much of these costs are the result of the psychological disability is unknown. However, those knowledgeable in workmen's compensation matters are well aware that when a workman is in poor mental health, it is the major factor in his disability.

It is rare for a workman to receive open and forthright mental health care in workmen's compensation. Yet some level of mental health care is probably needed in every instance of permanent disability. However, the number of permanently disabled workmen presently getting proper psychological counseling or psychiatric care as an obvious form of medical treatment in any given year is infinitesimal.

We are not an enlightened people in mental health care. Although we have stopped throwing our psychotics in dungeons, we still leave workmen in need of mental health care in the abyss. The bulk of the mental health care provided occupationally ill or injured workmen today is done so in a clandestine manner.

Most mental health care provided for workmen is given under the guise of treating the physical illness or injury. Additionally, it is often rendered by a treating physician who is too busy, or possibly insufficiently trained. This may be our most costly form of mental health care.

Most treating physicians are capable of handling minor emotional problems in their patients. Beyond these minor mental aberrations, the average physician needs consultative help for the patient.[11]

There are more seriously disabled workmen in existence because they were denied proper mental health care than for any other reason in workmen's compensation. This writer believes that this failure to recognize the mental health problems related to occupational illnesses and injuries is the single greatest cause of disability in most workmen manifesting moderately severe to severe disability.

Rehabilitation Benefits

What's going on in rehabilitation? The National Commission is calling for reform because the Commission believes rehabilitation is more than a mere gesture of social responsibility. It is economic wisdom.[2]

Rehabilitation is normally divisible into physical rehabilitation and vocational rehabilitation. Although they are separate in thought somewhat, they commonly intertwine when rehabilitating a workman.

Four basic problems exist in rehabilitation of workmen in workmen's compensation today: (1) lack of supervision; (2) lack of provision; (3) delayed rehabilitation; and (4) lack of incentive for the workman.

Presently twenty-two states have a rehabilitation division within their workmen's compensation agencies. However, according to the National Commission, this does not mean supervision of the workman's rehabilitation is adequate.

Physical rehabilitation is reportedly badly neglected in many states. Perhaps, in part, because an intermix of insurers, employers, and state agencies are seen providing physical rehabilitation to the workman.

The fact that physical rehabilitation is supplied to a workman by these different parties is no assurance that it is properly done, or that all workmen always receive it at the proper time. This is what underlies, in part, the National Commission's recommendations for closer supervision of rehabilitation by the workmen's compensation agency.

Even when a state's workmen's compensation agency is charged with the responsibility to rehabilitate the workman, it is no guarantee of a job well done. This occurred in an otherwise progressive workmen's compensation system in the West.

The Workmen's Compensation Board in Oregon operates a rehabilitation center for workmen. But the Board discovered it was not physically rehabilitating workmen enrolled in its center.

Many problems existed as reasons why the center was not providing proper physical rehabilitation. One reason was a fantastic delay in the referral of workmen to the center by treating physicians. The average workman being referred did not arrive at this center until eighteen months after his injury.

Additionally, the Board's inquiry into the matter revealed that the majority of workmen were undergoing treatment for too short a time in a very inadequate physical facility. The center was serving mainly as an evaluation center.

In two years Oregon did a turn-around in rehabilitation. A disability prevention program has been implemented. A new physi-

cal rehabilitation center is under construction. This new facility
will provide the very best available in physical restoration and
pre-work conditioning for Oregon's workmen.

Because supervision by most state agencies is lacking, what
passes for physical rehabilitation in the minds of some is merely
the laying on of heat and hands followed by slipshod exercises
of the injured part. Meanwhile, the remainder of the workman's
body is left to grow soft and incapable of work.

Furthermore, physical rehabilitation is limited, or may never
occur, in those states which limit the total dollar amounts spent
for medical care of the injured workman, or by those states limiting the duration of medical care. Almost unbelievable are those
states which flatly refuse to pay for care rendered by occupational
therapists or by rehabilitation centers.[3]

Vocational rehabilitation provided workmen appears to be even
less satisfactory than physical rehabilitation, in general. The National Commission does not believe that workmen's compensation
is doing an effective job.

Most vocational rehabilitation services provided by most states
are through vocational rehabilitation divisions funded by federal
money. But, the National Commission indicates there is little formal connection between vocational rehabilitation divisions and
the workmen's compensation agencies. As a result, too many
workmen needing vocational rehabilitation are not receiving this
service.

In addition, the National Commission finds vocational rehabilitation as now provided is not financed in the main by workmen's
compensation. The costs are being shifted outside the system.
The method of financing vocational training are a scrambled picture throughout the states. This, of course, is contrary to the basic
philosophy underlying workmen's compensation.

Beyond this, many workmen are discouraged from attempting
any vocational rehabilitation. The additional expenses to the workman require more financial assistance than is available to them.

It is difficult to determine what benefits are available to the
workman in each state. Take the matter of maintenance allowance: Maine pays the workman $35 per week for fifty-two weeks.

It is not clear if this $35 per week is in addition to, or in lieu of, temporary total disability compensation payments.[1]

Maintenance allowance does appear to be in addition to compensation income in some states. In other states it appears that it is not. However, limitations on either dollar amounts, or length of training time, or both, exist in thirty-six states.

There are also states in which it appears that a workman, in effect, pays for his own vocational rehabilitation out of deductions made from his permanent disability award. Under this circumstance, it is likely that welfare assistance is almost mandatory for a workman undergoing vocational rehabilitation.

In Summary

Many benefits for workmen exist in workmen's compensation. There are also many unnecessary deficiencies. It is time for reform, as the National Commission has stated. But any reform must be result-oriented for the workman, rather than merely a perpetuation of cost-conscious, piecemeal legislation which generates the adversary attitude.

THE ADVERSARY SETTING,
A NEGATIVE RESULT

A FAMILIAR RING

DIG A HAND INTO THE CLAIM FILES of any workmen's compensation insurer or agency. You are sure to come upon a claim file containing a workman's story strikingly similar to the following tale. It is not an uncommon one to those who daily review claims.

Between the covers of this file folder there is a tale of injury, expectation, gradual concern, anxiety, questions and fear, confusion, desperation, hope and failure, and finally, the fight for survival.

Our story is about Sam, the mill hand. Sam may even be a claimant of yours, or an employee, or your neighbor down the street. On a Thursday morning Sam strained his back while working at the mill. The following morning Sam had a deuce of a time getting dressed. By midafternoon he was sitting in his doctor's office.

After examining Sam, his doctor prescribed physical therapy treatments three times weekly for two weeks and a few pain pills. It was just a back strain, and the doctor's initial report of Sam's injury was on its way to the insurance company offices two days later.

Then, after two weeks of heat and a little light massage, the doctor saw Sam again. Sam's no better, both agree. Sam's to take two more weeks of treatment, and rest.

Sam's injury is a common one in workingmen his age. Sam's forty-two now, and he's worked hard most of his life. He was never much interested in school as a kid, and he quit after the tenth grade. After Sam married they had two nice kids and they managed to buy a good place. They and the bank.

Since his last visit to the Doc's office, Sam's been sitting around the house. He's been thinking some about his future. Bills are coming in. And not a soul from the mill has called or come by.

However, a check from the insurance company arrived in yesterday's mail. It is better than nothing, but it wasn't as much as he expected either, and that caused Sam a little more worry. Also he noticed a funny thing about his back; it's been a little worse since yesterday. Well, he'll soon be seeing Doc again.

"Two more weeks of therapy," Doc says when Sam goes back. "Then we'll see how things are." After Doc tells Sam this, Sam decides he'd best call the mill and talk to his foreman. Sam knew they'd put another guy on his job. Probably Marvin. That Marvin's been wanting Sam's job for a long time.

Sam's job is one of the easier ones at the mill and he's been thinking maybe he ought to try going back to work before they give the job to Marvin for good. Still, the way his back's been hurting, he isn't sure he'd last out the day. One thing is certain, Sam decides. He couldn't work a tougher job, not with this back aching worse like it's been of late.

Meanwhile, at the insurance company's offices, Sam's claim is already established. The initial reports are in. A request for a progress report from Sam's doctor has gone out. The first compensation check to Sam went in the mail several days ago. They're in compliance. Everything necessary is being done. The claim's manager is pleased with their routine efficiency.

On Sam's next visit to the Doc, things still aren't any better. Sam's complaining more and Doc's looking more concerned. Obviously, hospitalization and traction is the answer for Sam's problem since the heat treatments aren't helping, according to Doc.

Ten days on your back in a hospital is a grind, Sam concludes. He's been lying there mulling over the problems a strained back can bring a man. Adding to his worries is that bill collector hounding the missus, and, his younger kid came down with strep throat yesterday. More bills!

Sam tries thinking about some good things. The insurance company is paying his doctor and hospital bills. He owns almost half his house and he can go back to the mill when he does get better. Things aren't too bad. But it seems he can't stop thinking more and more about his troubles either.

What if his back doesn't get better? What if he doesn't get his old job back? How's his back going to hold up on another job? Another job sure might be too tough.

Can that bill collector grab his compensation check? He and the missus sure would be in the soup then. That compensation check isn't going near as far as his pay check, and the missus is worrying, he can tell.

Anyway, the check's still better than nothing coming in, Sam decides. His missus is good at buying groceries and still paying something on them bills. Not all of them every time, but some of the bills. What's worrying Sam more is the traction isn't doing much good. But them shots and the pain pills sure do ease the pain. And another good thing about them shots, they take away some of a man's worries.

Well, he's going home in another day or two. It's been two weeks since he came to the hospital. Doc's looked a little puzzled the last few days 'cause the traction ain't helping. That's why Doc asked the new doctor to see Sam. The new doc's a bone specialist. What's a myelogram?

Do they ever make a mistake sticking a needle in a guy's back? Who was the guy at the mill they left crippled sticking a needle in his back bone? Wonder if it really happened? Sam winced just then with another of those grabbing kind of pains.

Well, the myelogram didn't show anything. Shucks, it isn't so bad having that needle stuck in your back, Sam's thinking. Besides, his back is going to get better. That's what the new doc decided.

He's going home tomorrow. Another month or so around the house should make his back a lot better. Bad news, though. They're giving his job at the mill to Marvin. The foreman called and told Sam's wife so last night. They couldn't be expected to keep a job for him forever. They might have a job for him though when his back heals. That is, "if business picks up."

It's more than four weeks now since he left the hospital and the new doc isn't sure if Sam can ever go back to the mill.

The new doc also mentioned something about X-rays not always showing a "disc." He also mentioned that maybe Sam ought to see one of those government agencies. They might be interested in training Sam in doing something else. Then, another thing. The new doc was beginning to think Sam might need surgery yet.

The insurance company finally answered Sam's letter. Yes, sometimes they trained a man for a new job. But they couldn't consider such a thing unless his doctor recommended vocational training; and then they would have to review his case.

Well, that's it. He's going to have an operation. They're going to take out his disc. A disc is something that pushes out between your back bones and pinches your nerves.

"Not all discs are typical. Conservative treatment isn't helping. Surgery is the answer." That's what the new doc told Sam.

Besides, the pain pills were the only thing helping Sam, of late. But it was taking more them to keep the pain down.

Six months ago Sam sure never thought he'd be talking to a real estate salesman about selling the house. Well, they can rent a place until he gets back on his feet. An apartment, maybe. The money from the house will pay some of the bills they're owing. The rest of the stuff, the stores will just have to take back.

After the surgery Sam seemed to feel better for awhile. But lately his back's been paining some. Adding to this worry, when Sam called the mill he found, "Business wasn't getting any better."

The foreman didn't have any jobs Sam could handle. Well, anyway, he was still getting the compensation checks. A funny thing happened, though. A man from the insurance called yesterday and told him the company thought he ought to be getting better.

Nine months is a long time to sit around doing nothing. Besides that, the other day the new doc decided there wasn't anything more he could do for Sam's back. However, Sam's back is better than it was before he operated on it, the new doc told Sam.

For Sam's sake, the new doc was telling the insurance company they ought to teach Sam something else now that they could close his claim. The new doc also thought Sam ought to get some kind of a settlement. And that was good, 'cause it would give Sam something to live on while they taught him something else.

Sam wanted to know if closing his case meant he didn't have anything else coming. The new doc shrugged. That was something legal, and doctors didn't understand all such things.

Sam's counselor at Vocational Rehab seemed nice, but she talked a lot. Some of it was funny stuff he didn't understand. However, after the tests they'd know how to help him.

The guy who gave Sam the tests asked some real nosey questions and it all made Sam feel strange and nervous. They must think he was sick in the head. Funny thing, those tests even made his back hurt worse and he hadn't done nothing but sit there in the chair.

Afterward, his counselor told Sam he was very limited in his education, and he told her that he knew that. He'd gone to high school two years; but, "It was just funnin' around." If that was what the test was for, he could have saved her all the trouble, Sam told her. And she gave him a funny smile.

Anyway, she was right. He didn't have a knack for a desk job. And bending over a car fender working on motors wouldn't be good for his back. Too, Sam didn't think he'd be much interested in cutting hair. But he agreed to come back after she'd analyzed his case more.

A couple of weeks later, after the counselor called, Sam went to her office. She said he would make a good welder, and Sam agreed to try it. But he didn't hanker to it too well. Sam already knew four guys who could weld, and they weren't working. He never mentioned to her that his back was getting worse since that last visit.

When his "reward" check came, Sam knew he'd never forget that day. His back started killing him soon as he saw how much reward they gave him. Even his wife wasn't sure what all the words on the piece of paper meant. All it meant to them was, the company had closed his case, and he was 15 per cent disabled.

He'll be hard put to buy them pain pills now. Bills were already taking up all the compensation check. Sam groaned. Then for a bit he thought he was going to be sick at his stomach. The pain in his back started grabbing worse than ever. He sure wished he could see Doc. But who'd pay the doctor bills now?

After resting on the bed a spell, his back eased off hurting and Sam thought more about what he'd been thinking lately. Maybe he'd ought to see that guy at the mill. The one the insurance company screwed that time. That guy got a lawyer, and made them pay good.

At the lawyer's office Sam told the lawyer he couldn't see how he could ever work again. Not with his back killing him all the time. He was crippled for life. "Just a basket case" . . . and the lawyer slowly nodded his head.

THE MOCKINGBIRD

The adversary process is amazingly well and outgrowing its paternalism in workmen's compensation. This is possible only because the system's design nurtures adversity. Although the preceding story of a workman's claim is a fictionalized version, it smacks of an ordinary commonness and certain truths found all too often in workmen's claims. The elements of adversity strewn through this story ring familiar to those knowledgeable in workmen's compensation affairs.

Forty-five million of us suffered disabling work injuries in 1971.[3] What we don't know is how many suffered disastrous personal losses as a result. But, in spite of the enormity of our problems in workmen's compensation, they remain a whisper in the public debate when compared to our louder vocalizations on matters such as minimum wage laws, auto insurance, pollution, abortions, and even the Watergate affair.

Poorly constructed social legislation deludes, deprives, and destroys those it purports to help. Excepting those workmen already experienced in the frustration and the struggle to survive after a disabling injury, we all think of workmen's compensation in a generally abstract manner; as though we ourselves, will never be touched by a disabling injury.

Our apathy toward and our ignorance of the system permits the continuation of this insufficient system. According to the National Commission, various interest groups now thwart most efforts made to meet many standards developed to correct the serious defects existent in the system.

Sixteen standards already exist based upon the recommendation of national organizations knowledgeable in workmen's compensation. These standards were set forth by the Department of Labor in 1959. Today, after fifteen years, the average state meets only eight of these standards. And there are reasons.

Employer groups, unions, lawyers, and insurers, all ply varying degrees of pressure upon state legislators to protect their interests.[4] The legislators, in turn, find the workman's compensation system complex and confusing.[5] In addition, they can ill afford the time it takes to understand the intricacies of the system.

As a result, state legislators are often dissuaded from effecting reform. We find legislators fearing that reform will bring overwhelming costs that will drive business out of their states. As a result, attitudes established in the minds of the legislators cause them then to burden their state with (1) limiting legislation in the workmen's compensation acts; (2) continuance of passive or weak administration of the statute; and (3) perpetuation of the adversary setting.

As an example: The legislature restructured the entire workmen's act in Oregon during the 1965 legislative session. Intensive negotiations between labor and management and other interest groups took place. Together they strove to design a self-administrating law which hopefully would nearly nullify the adversary process.

Under this revised Oregon Workmen's Compensation Act the legislature eliminated jury trials. They thought the new act would also generate fewer appeals. They believed few appeals would ever reach the courts because they established two levels of appeal within the Workmen's Compensation Board.

Although the revised Oregon law may be more equitable than the previous one, it did not eradicate the adversary process. Additionally, the 1973 legislative session saw a bill introduced to reinstate jury trials for workmen's compensation claims. Although this bill was finally defeated, be assured the pressure upon the legislators by those proposing this return to jury trials was strong.

The Oregon experience is another example added to the long history of a zealous struggle to protect self-interests and rights in workmen's compensation. Meanwhile, the hopes of meeting the needs of the occupationally ill or injured workman in an equitable manner remain mostly hopes.

Matters are obviously askew in Oregon. Were they not, why else the effort to move backwards toward the more time-consuming, more expensive jury trial system of appeals?

Oregon is considered one of the more progressive states in workmen's compensation affairs, but the adversary process is in full bloom there. Oregon's belief that its new act would markedly reduce litigation is not borne out.

Table II shows the annual increase in the number of appeals in Oregon's workmen's compensation system for the fiscal years 1967-72.

TABLE II

APPEALS OF OREGON WORKMEN'S COMPENSATION CLAIMANTS

Fiscal Year	T. C. F.*	Hearing Officer	Board Review	Circuit Court	Supreme Court
1967–68	103,938	1978	410	155	11
1968–69	117,223	2189	447	223	19
1969–70	113,528	2537	500	273	21
1970–71	109,052	2697	577	297	25
1971–72	118,825	3301	535	295	9
% increase or decrease	13%	40%	23%	47%	−36%

* Total numbers occupational injury and disease claims filed.

As it is with the call of a mockingbird, the adversary process continues to mimic the best of legislative efforts to eradicate litigation in workmen's compensation. Litigation is, and will be, a continuing burden borne by the system so long as attempts to reform are limited purely to those that are legislative in nature.

BENEATH THE CHOP

Determining the enormity of the adversary process in workmen's compensation is probably impossible. The National Commission met with frustration in its effort to determine these facts. They found that information on contested cases is far from complete, in general.

The percentage of claims contested among the states range from a low of less than 5 per cent in eleven states to more than

50 per cent in one state, according to the National Commission. In another twenty-five states they found it impossible to determine any facts.[2]

Fact-keeping in workmen's compensation is a shambles. In general, it is unbelievable that so little information depicting the results and the effectiveness of such a large system is gathered. Also, no standardization of fact gathering exists. This makes comparative analysis of facts from state to state nearly impossible and a lesson in frustration.

No one knows for certain the involvement of the legal profession in workmen's compensation claims. However, one needs to spend but little time talking with the professionals in workmen's compensation to realize only the tip of the iceberg is showing.

Florida is a state doing one of the better jobs of keeping understandable facts. Florida's annual reports[6] gives us some hint of the growing litigation occurring in workmen's compensation as seen in Table III.

TABLE III
STATE OF FLORIDA
CLAIMS APPEALED

Year	T. C. F.[1]	J. I. C.[2]	N. O. A.[3]
1968	245,469	11,868	7,562
1971	290,046	24,020	8,373

[1] T. C. F. – Total number occupational claims filed
[2] J. I. C. – Judges of industrial claims
[3] N. O. A. – Number of awards granted

Between the years 1968 and 1971 Florida shows an 18 per cent increase in the total number of workmen's compensation claims filed annually. However, we see more than a 100 per cent increase in the number of appeals.

Also hidden in the root of this iceberg is the matter of compromise and release settlements. Most facts relating to this form of settlement and claim closure are unavailable for scrutiny.

The National Commission found it impossible to determine accurately how frequently compromise and release settlements are used in workmen's compensation. As a percentage of all cash benefits paid to workmen, the Commission found the settlements made range from 5 per cent in six states to more than 50 per cent in one state.[2]

THE HOPE

Underlying each occupational illness or injury claim is an individual having certain expectations of the workmen's compensation system. We all have a general awareness that the system provides certain benefits. We each hope these benefits are adequate for our needs as we recuperate.

We each expect that certain basic benefits be provided. First of all, we naturally expect to receive prompt and good quality medical treatment. In addition, we anticipate being compensated for time lost from work. And we hope these cash benefits provided are adequate to support our families during the time we can't earn a wage.

In addition, we each have differing expectations of the system beyond these two most basic needs. For some, the injury or illness leaves them permanently crippled and they expect additional payment for their crippling.

For others, the nature of the injury or illness means they can never return to the old job; and they hope the system will train them to work again in another job or trade. Regardless of individual expectations, we all hope that the benefits provided by the workmen's compensation system meet our needs according to our specific predicament.

But when the benefits we expect do not come to us as expected, it is then that we find ourselves reacting to the system. We begin to see the system as our adversary. And this is not unnatural. In spite of the lawful assumption that no fault exists in event of an occupational illness or injury, there is still a fundamental feeling of fault in the one made ill or injured by their work and the wording of the law does not necessarily erase those thoughts of fault from our minds.

Where thoughts of fault exist, the elements of blame and anger are likely to follow. Once these are manifest, we are prone to seek retribution by engaging our opponent in litigation.

We see our opponents within a workmen's compensation system as one of several: The employer, the insurer, the physician, or the state agency; or, the opponent may even become the whole of society in a workman's mind.

THE TURN-OFFS

Those things which turn off a claimant-workman and send him scurrying to an attorney are highly variable. Each instance of litigation is an example in the individuality of these variables. However, certain common factors frequently exist that gives us things to consider as common turn-offs.

Rights of the Workman

Ignorance raises questions, doubts and suspicions, fears, misbelief and confusion, frustration and anger and adversity. Within the respective workmen's compensation laws, the rights of the workman are prescribed by the law. Making these rights clear to all parties and all persons at the time a workmen's compensation claim arises is a common omission leading to litigation.

The average workman has no clear idea of his specific rights and those benefits provided for him by the workmen's compensation system. Most amazing are the continual examples seen wherein the workman is totally ignorant of his rights, benefits, and obligations. This does not mean the workmen's compensation system is not making an effort to inform workmen of their rights and benefits. It simply means their efforts to inform are generally ineffective.

Efforts to inform probably fail because of the manner in which most claims are processed. The system's tendency is to ritualize claims processing for purposes of greater efficiency and the lowering of costs. As a result, the system's claims processing dehumanizes the claim, the injury, and the workman.

A recent example showing a result of these impersonal claims management methods came to light in the records in one of the more progressive states. The workmen's compensation agency in this state publishes and mails to each claimant a well-designed pamphlet explaining the rights of occupationally ill or injured workmen.

Each claimant workman is mailed a pamphlet at the time the claim is made. It is a policy ardently adhered to by the agency. And it is sincerely believed by the agency that workmen in their state are being informed of their rights. However, they missed one workman, and maybe more.

Somehow, their pamphlet failed in its mission in this specific instance of a seriously injured workman. Fortunately for him, he was not totally ignored. Quite by accident he was discovered still in the hospital some four months later by a member of the agency.

Still hospitalized, and still totally ignorant of his benefits and rights, this workman sorely needed advice. He was filled with questions and worries, among which was the matter of medical bills.

Unaware of workmen's compensation coverage or benefits, he visualized himself in deep financial troubles. He apparently never recognized the source of his compensation income checks and saw them as an act of a benevolent employer. This, he believed, would cease with time.

Although this man is now aware of his rights and benefits, it does not follow that the problem is corrected. He may harbor feelings of resentment toward a system that has let him stew and fret these many months. If so, the slightest error later in the management of his claim may trigger litigation.

Likely, some readers may view this example as an isolated happening in workmen's compensation, that the problems arising out of the uninformed or misinformed workman are irrelevant. This is not the viewpoint of the National Commission, however.

The National Commission's report sets down six obligations administration must face if we hope to create an effective delivery system in workmen's compensation.[2] One of these six obligations is to clearly inform all workmen of their rights, benefits, and obligations under workmen's compensation law.

More Rights, Wrongs, and Results

The door leading subsequently to the adversary attitude and litigation is easily opened by the initial events occurring in the workman's claim. All parties to the workman's claim have certain rights and obligations to meet; and, herein, hang some rights and wrongs.

For a workman to establish a compensable workmen's compensation claim, it usually depends upon the employer making an initial report to the insurer in most states. Failure of the employer to report promptly, or refusal of the employer to report to the insurer, is a denial of the workman's claim.

Such a circumstance then leaves a workman faced with a fight for his rights from the beginning. This is especially so if there is outright denial by the employer of a work-related injury or illness. The workman's only choice is to appeal. In most instances, this requires the services of an attorney to make the appeal, whereupon the adversary attitude is then well set in this workman's mind.

The insurer's decision to accept or deny the workman's claim, or to pay income benefits, are also partially dependent upon the treating physician's report. A physician's failure to render a prompt and proper report to the insurer can certainly cause the insurer to err and thus create an adversary attitude in the claimant workman.

Many physicians fail to recognize that their obligation to the workman extends beyond medical treatment in workmen's compensation. Too many physicians do not understand the impact poor medical reporting has upon the workman's life. For example:

The timeliness of the initial medical reports submitted to insurers by physicians in Oregon during 1972 reveals a general laxity of Oregon's physicians in assuming a basic responsibility to the workman, and to the insurer under workmen's compensation law. Less than 18 per cent of Oregon's physicians submit their initial medical reports on time.

Excluding holidays and weekends, Oregon physicians are required to make the initial report of injury and treatment to the insurer within seventy-two hours. This is done in order to give the insurer ample notice of a time-loss illness or injury. In turn, the insurer is required to make the first time-loss payment to the workman on the fourteenth day.

Presently, 43 per cent of Oregon's physicians are not making their first report to Oregon insurers by that fourteenth day when the time-loss payment is due the workman. This places the insurers in a trap because they must face one of two decisions:

1. Risk overpayment of time loss to the workman, with little chance of recovering the overpayment;
2. Risk late payment to the workman whereupon the insurer is penalized.

In addition, an audit to determine the timeliness of first payment to the workman by Oregon insurers is also reported annually. Currently, the first compensation income payment to the workman is late 15 per cent of the time.[7]

Late payment of the first compensation income check has an impact upon the workman when one reads the report of Page and O'Brien.[8] Their report reveals 68 per cent of the workmen disabled beyond fourteen days are in financial trouble, and this alone is enough to stress any workman.

That stressful events such as this, and others, lead to the retention of an attorney by the workman is reported in the study of Leavitt, et al.[9]

Leavitt's study shows certain stressful events in the areas of claims management, medical care, or the return to work pattern are primary causes of litigation in workmen's compensation.

Looking first to claims management, the manner in which benefits are provided to the workman is determinate of the reaction in the workman. Leavitt found the commonest events in claims management that precipitate litigation to be:

1. Rejection of the workman's claim for whatever reason;
2. A request to the workman for repayment of compensation income overpayment;
3. Termination of temporary total disability benefits;
4. Certain attempts to close the claim; and
5. Disputes over permanent disability.

An example showing how the management of a specific workman's claim by an insurer triggered the adversary attitude leading to litigation is seen in the following facts extracted from an actual claim file:

This workman fell from a tractor-trailer rig on February 11, 1972. He injured his back in the fall and promptly saw his doctor. Ten days later he was referred to a consulting orthopedist. Diagnosis: acute lumbar strain.

According to the orthopedist, this man's back strain was so severe he could not resume working. Thus, the insurer began paying the workman $85 per week for temporary total disability.

In mid-April the employer's report of the workman's gross earnings caused the insurer to conclude that they were overpaying the workman. Without explanation, the insurer immediately reduced the disability payments to $58.76 per week. And that was a mistake! The workman went straight to an attorney.

The attorney promptly wrote the insurer stating his intentions to file an appeal requesting restitution of the disability income due his client, and he intended to request that the insurer be assessed a 25 per cent penalty fee because of their arbitrary actions.

In addition, the attorney also requested that the insurer place this workman into a vocational training program as previously recommended by the orthopedist.

Upon receipt of the attorney's letter, the insurer promptly increased the workman's disability payments back to the level of $85 per week. However, they ignored the request that this workman be vocationally retrained.

Medical treatment continued until June 1972. Then the orthopedist, in a very short report giving little or no medical information, recommended claim closure. The claim was immediately closed without any award for permanent disability.

Shortly after claim closure, the orthopedist then asked that the insurer to re-open the claim and that the workman be referred to a special back clinic for examination. The orthopedist noted in his request to the insurer that the workman had attempted to return to work but could no longer drive a truck because of his severe back pain.

Presently this claim is in litigation. The argument now is over the degree of permanent disability caused by this injury. It is little wonder. The ingredients leading to litigation are certainly present in this claim. The four specific ingredients present in this workman's claim which Leavitt notes as stressful events leading to litigation are:

1. Overt anger arising in the workman as a result of indifferent claims processing in the insurer's office;
2. A treating physician unsure of the cause of the workman's back pain;
3. A claimant who failed in attempts to work; and
4. A dispute over disability.

It is unlikely that a single incident ordinarily causes a workman to seek an attorney. Such action is probably most often the result of the cumulative effects of stressful events occurring during the course of recovery. However, Leavitt notes that there is usually a stressful event occurring within two weeks of the date a workman retains an attorney.

As a result of the California study, Webb[10] considers disputes over the extent of disability as the major cause of litigation in workmen's compensation in California.

According to Webb, disability ratings below 25 per cent permanent disability are considered minimal disability by insurers. Webb notes that although this is looked upon as minimal disability by the insurers, it represents injury and pain and a threat to the workman's future. Webb believes the insurer's failure to pay more heed to a workman whom they assume has a minimal permanent disability drives that workman to an attorney.

TABLE IV
HEARING OFFICER STATISTICS, 1971

Issues	% of total hearings
Extent of permanent disability	57%
Denied Claims	18%
Aggravation of pre-existing injury	7%
Extention of temporary total disability	6%
	88%
Workmen's Compensation Board, State of Oregon	

Oregon's data support Webb's findings that disputes arising over the extent of permanent disability are the major cause of litigation. The workman's first step in the appeals process in Oregon is an appeal before the Hearing Officer. The following issues and their percentages show those issues accounting for 88 per cent of the appeals made by Oregon's workmen.

Except for those workmen contending permanent total disability in Oregon, the average workman appealing for an increase in the permanent partial disability award was also previously granted a small award for permanent disability.

The second major cause of litigation in California is related to the subjective complaints of the workman. Failure to give reasonable credibility to the workman's complaints prompts the workman to seek an attorney.

Beals and Hickman's study lends credence to the significance of the workman's subjective complaints.[11] Their findings indicate that workmen having recent injuries display moderately severe depression, but only mild hypochondriasis, and mild hysteria. If they remain disabled, a reversal of these neurotic reactions occurs.

The findings of this reversal from moderately severe depression in the early stages of injury to that of moderately severe hypochondriasis and hysteria with the passage of time is significant.

The hypochondriacal-hysterical person is the complainer and the seeker of treatment. These are persons who are both aggravated and aggravating. They focus their attention upon their physical, or somatic, feelings. Each twinge of pain, or even an itch, is magnified out of proportion. They are not malingerers, however.

To ignore their complaints, to view them as malingerers, or to persist in treating their physical complaints worsens the hypochondriacal-hysteria problem. Insurers, agency people, employers, and even physicians are prone to seek an answer to this particular workman by looking only for physical damage. But, if no objective findings exist, they tend to overlook the workman's subjective complaints.

The tendency is to give less and less credence to the injured workman who complains more and more. Most of us tend to turn-off the man or woman who bemoans their predicament.

And, in doing so, we force the workman to seek a sympathetic ear. Claimant's attorneys are the ones among us having a sympathetic ear for the workman's subjective complaints.

In a sense, litigation is therapeutic for the workman having a great number of subjective complaints. It forces others to acknowledge his complaints. And, it permits him to vent this hostility which we create in him by ignoring his complaints. However, it also creates some wrongs.

The wrong arising out of the adversary process when it involves a workman suffering hypochondriasis or hysteria is the disabling effect. Should this workman prevail in the courts, he has then proven to society his disability. However, if the courts deny that this workman's disability exists, his hostility toward society is likely aggravated because he remains convinced he suffers.

Insurers, employers, and workmen's compensation agencies are also prone to close the claim of workmen having considerable subjective complaints, or of those claiming greater disability than is substantiated by physical findings.

However, manifest hypochondriasis or hysteria is not lessened by premature claim closure. Such attempts serve only to aggravate the condition. Hypochondriasis or hysteria require the same proper medical care as does fractured femur or a herniated disc.

Moving on into medical care and the stressful events noted by Leavitt which lead to the retention of an attorney, we find them to be:[9]

1. Hospitalization,
2. Authorization for surgery,
3. Conflicting opinions between physicians regarding proper treatment, and
4. Persistent symptoms and disability after treatment.

Also relevant to these specific events is the passage of time. Leavitt finds the decision to hospitalize the workman or requests asking for authorization to do surgery came later in those claims where the workman subsequently retained an attorney. And, once hospitalized, these workmen remained hospitalized more than twice as long as the national average for the specific type of injury. The passage of time is also involved when differences occur between physicians as to what they consider proper treatment.

Leavitt also found in the group of workmen retaining an attorney that they were examined, or treated, on an average by 5.2 physicians. This contrasts to the group of workmen who did not retain an attorney and who were treated on an average by 1.3-2.0 physicians.[9]

In spite of the fact that all workmen retaining an attorney received more examinations and more treatment, their symptoms and disability persisted. What Leavitt seems to bring out by all these findings goes back to the disproportionate relationship between subjective symptoms and objective findings in workmen prone to developing an adversary attitude.

The matters of hypochondriasis, hysteria, depression, as seen in the group of workmen studied by Beals and Hickman then relate to Leavitt's findings. They are significant findings in that these workmen do not respond to the usual physical treatment. And, in general, there is a tendency for the average treating physician to overlook emotional disturbances present in these workmen.

Treating physicians are generally oriented to the physical, or somatic symptoms, of the workman. As a result, treatment is directed toward correction of the physical injury. Thus we find the majority of physicians not giving enough consideration to the workman's psychological needs and ills.

Thirty-six per cent of the workmen in the group Leavitt found likely to retain attorneys were also noted to have psychological problems by their treating physicians. Although their physicians recorded the presence of emotional disturbances in the workman in their records, not one referred the workman for proper mental health care. All referrals were made to other physicians equally as oriented toward treatment of physical ills as was the treating physician.

A physician who does not have an ear for his patient's, or workman's, subjective complaints is apt to be led into a trap. Beals and Hickman found that as the patient's emotional distress increased, so did the resulting number of surgical procedures in the patient.[11] And, they become treatment failures in the end.

Any physician who persists in baking-out, massaging-out, or cutting-out the subjective complaints of a workman having this sort of psychopathological distress always fails to help the workman. Ultimately, the physician gives up, or becomes confused, or grows weary, or develops an adversary attitude toward this workman who resists the physician's curative skills. The physician then either refers the patient to another physician or steadfastly maintains the workman has ability to work and recommends claim closure.

In this circumstance a workman's response to his physician is no different from his response to the insurer or employer who ignores his complaints. The workman seeks an attorney and the attorney has an ear for the workman.

The inability of a workman to resume working is nearly always a major area of contention. It is not surprising that an adversary attitude develops in any workman failing in his efforts to go back to work. However, much of a workman's attitude depends upon his response to those who would have him work.

In general, many workmen failing in their efforts to return to work find themselves suspect. Our responses as employers, insurers, physicians, or even as neighbors, toward the workman slow in returning to work is too often skepticism. And, unless there is absolute proof of substantial physical crippling of the workman, our attitudes evolve to dogmatism.

The credibility of subjective symptoms in certain types of physical injury is very difficult to implant in the minds of others. It is almost as though it is a matter of an out-of-sight, out-of-mind type of thing. If the workman's physical damage is not obvious to us, then neither are the symptoms.

An example of physical injury in which the damage and the cause of the symptoms are not always easily apparent is seen in a workman having had a heart attack, the workman in whom one of the heart's arteries has plugged and severely damaged a part of the heart's muscle—the myocardial infarction.

Any workman having a myocardial infarction may well suffer

from chest pain, weakness, shortness of breath, and abject fear when healed. But who among us can see this workman's physical damage?

However, setting aside those subjective symptoms directly attributable to the physically damaged heart muscle, the matter of psychological damage still remains in this workman. Who among us feels or sees the abject fear besetting this workman's mind?

In workmen's compensation it ultimately becomes a matter of which it is that is now crippling this workman, his mind or his heart. Then, if the physician's statements do not support the workman's subjective symptoms with provable evidence of sufficient heart damage, the burden of proving his disabled state quickly becomes that of the workman. Thus, he is placed in the adversary role.

This adversary attitude relating to a workman's workability is also a finding in Leavitt's study. A stressful event occurring in the worklife is one of the reasons likely to cause the workman to seek an attorney. Two main patterns of stress are noted by Leavitt:[9]

1. The workman continues to work after injury until his disability becomes very obvious;
2. The workman makes repeated attempts to resume working, but fails with each effort.

It may be that the psychological make-up of the workman who persists in working after his injury ultimately causes him to retain an attorney. He likely suffers the *strong man syndrome* noted in Beals and Hickman's study.[11]

Beals and Hickman found the group most severely disabled was also the group of workmen manifesting the most masculinity. It is unacceptable to these workmen to be less than strong, whole men.

If such a workman cannot be the masculine image he has of himself, it is equally important to him to prove he is sorely disabled. Beals and Hickman found these workmen had a great need within themselves to have their physicians recognize their disability. It is likely this same need for recognition carries over to employers and fellow workmen.

Leavitt finds that many workmen making repeated efforts to return to work are authorized to do so by the treating physician.[9] However, the physician is misjudging the workman's ability to work, and by this mechanism the elements of the adversary attitude develop.

Of course the attitude of others also enters into a workman's ability to return to work. The willingness or ability of the employer to employ the workman has direct bearing. The attitude of a workman's labor union is also a factor in the workman's ultimate attitude, and whether or not he successfully returns to work or retains an attorney.

When the survival needs of a workman are challenged by a threat to his work-life, it produces realistic fear and outright hostility, according to Meineker.[12] This type of reaction should not come as a surprise to anyone if we think for a moment about our reaction to any threat to our own survival.

However, the workmen's compensation system, by its general design, does challenge the survival needs of the workman. We have laws devised by men intending to insure the workman reasonably well against poverty arising out of an occupational illness or injury. At the same time, the law intends to protect the employer against bankrupting liability suits. But, under scrutiny, the entire system appears to lean more toward insuring the workman a dole.

We see a system where administration ranges from practically nothing to a fair performance. We see a situation wherein the parties to the workmen's compensation law function as isolated, or even as autonomous units, each doing its own thing for the workman in the majority of states.

As a result, the serial effect predominates in workmen's compensation. The workman receives what the law prescribes—bit by everlasting, antagonizing bit.

Workmen are frustrated and confused in many ways as recipients of the care and the protection provided. In effect, most workmen's compensation laws today are limiting laws. These laws limit employer liability while hamstringing the administrations of these laws.

As evident by testimony, the majority of the states have

limited income benefits which are below the poverty level. Additionally, the medical care provided varies in quality and quantity.

A general assumption exists that the physical rehabilitation given as part of the medical care is of high caliber. But there is reason to question the timeliness and the grade of physical rehabilitation commonly provided.

Vocational rehabilitation or training as provided is not meeting the needs of the workman nor the employer. The National Commission is very clear in its criticism of the effectiveness of vocational rehabilitation provided workmen in most states.

For the most part, it appears that a workman needing vocational rehabilitation is caught in the middle of interagency dysfunction. As a result, it comes as too little, too late, too costly, and too confusing for the workman.

Excepting middle-sized and larger corporations or companies, most employers find workmen's compensation laws confusing. This means that a large majority of employers in the nation harbor beliefs and misbeliefs that cause them to view the injured employee as a costly threat.

With the majority of employers not being knowledgeable in workmen's compensation laws it remains the domain of the insurers. The profit motive is always strong within the structure of the insuring companies, and even within some agencies. Actions or inactions of the insurer are always with the cost in view, and that is as it should be.

While costs must be kept in mind, too great a cost consciousness causes the insurer to concentrate upon meeting the minimum requirements under the law. Merely meeting the minimums leaves more to be desired, and this leads to more contention.

Contention is dispute and the presence of an adversary attitude. This in turn requires the services of the attorney in workmen's compensation because the adversary setting is usually quite formal.

The competent attorney pleads the cause of his client in a manner to obtain the maximum benefit. And, as we have seen, the bulk of disputes between the claimant-workman and the insurer are over the extent of permanent disability. But what does the workman win in winning?

How is it a workman benefits by being declared more disabled? A few more dollars? Greater future security? A more productive life? Or, less opportunity in the labor market and the world of work?

BUTTON, BUTTON, WHO'S GOT THE ... ?

The great value of the National Commission report is in its highlighting of our abysmal ignorance of the results arising out of a multi-billion dollar annual expenditure in workmen's compensation. Of the facts kept, most are with an eye to costs, not results of the expenditures. For example: no facts exist that give any idea of the number of workmen who ultimately join the welfare ranks as a result of their occupational injury or illness. Nor is it known how many workmen live a lesser life because they prevailed in the adversary process. Some knowledge of the expenditures in the adversary process exists, but there is no factual knowledge of the true costs resulting in the litigation so prevalent in workmen's compensation.

Comprehensive fact keeping wherein comparative interstate results can be studied is not a fact. Nor is any obvious research underway to determine the facts. The strongest search for informative facts to date is that of the National Commission.

The likely response to the National Commission report is reaction. We haven't a base of knowledge needed to act otherwise. The present system is incapable of effecting any changes in workmen's compensation that are likely to reduce litigation or minimize the adversary setting.

The National Commission found that a comprehensive appraisal of the direct costs of litigation is presently impossible. Isn't this amazing? It is incredible when we consider the fact that the National Commission believes the legal fees charged by both plaintiff and defendent lawyers are a large part of the costs in workmen's compensation.

The National Commission was unable to estimate the costs in 47 states. In two states, the costs appeared to be between 5%-10% of the benefit payments.[2] There are other hints and glimpses of the costs, but it is a piecemeal effort.

Florida offers some indication of the costs of legal fees, in part.[6] As shown in Table V, between 1967-1970, we see an increase of 48 per cent in claimants' lawyers' fees awarded as a result of litigation in that state.

TABLE V

ATTORNEY FEES AWARDED BY JUDGES OF INDUSTRIAL CLAIMS
FLORIDA

	1967	1970
Attorney Fees Awarded by Judges of Industrial Claims	$4,093,600	$6,082,183
Number of Awards	7,451	9,023
Average Award	$ 549	$ 674
Attorneys' Fees Awarded by Industrial Relations Committee	30,435	40,540
Total Attorney Fees Awarded	$4,124,035	$6,122,723

These dollars paid to the plaintiff lawyers in Florida are, in all probability, less than one-half the total direct costs of litigation. No indication is given of the earnings of the defense lawyers. In addition, there are also the administrative costs of maintaining an appeals system in Florida.

Based upon the reported fees paid to claimant attorneys in 1970, it appears reasonable to estimate that the total costs of litigation in Florida amounted to fifteen million dollars. This is sobering when we find Florida paying compensation income benefits amounting to only 22.4 million dollars to permanently disabled workmen in 1970.

It becomes more flabbergasting when we compare the costs of medical benefits for the permanently disabled to costs of litigation. In 1970 Florida paid less for medical benefits than it did for the litigious hassle. Medical care costs amounted to 13.1 million dollars as opposed to the estimated fifteen million dollars in legal costs for the workman.

If Florida is indicative, sizable sums are drained from the system by the adversary process. If the adversary process did not exist in Florida, there could be a 67 per cent increase in income benefits paid to the permanently disabled workman. Medical benefits could be doubled. Or, the savings could pay for a very comprehensive rehabilitation program that could bring a great many returns to the state of Florida.

Florida is probably indicative of the litigious goings-on in the majority of the states. In the California study it is reported 25 per cent of the total number of occupational disease and injury claimants consume 90 per cent of the dollars spent. However, more than 50 per cent of those dollars spent are on attorney fees and other costs of the adversary process.

The National Commission indicates it feels that the manner in which claimant's attorneys are commonly paid is also an incentive to litigation. It is common practice for the claimant's attorney to receive a percentage of the workman's award as a fee for service.

Aside from the manner claimants' attorneys are paid, statistical data gathered in the state of Oregon indicates other reasons claimant workmen are encouraged to litigate.

The Fisher report [13] reveals that any workman in Oregon granted a permanent partial disability award should contest the decision from the monetary gain viewpoint.

Looking at the monetary gain made by the workman in appealing his claim before hearing officers in Oregon in the year 1971, we see the average workman making an appeal had a 75 per cent chance of gaining $2,640.

However, the workman doesn't get it all. Claimant's attorneys are granted a fee of up to 25 per cent or $1,500, whichever is the lesser, out of a workman's increased award. This leaves an average workman with a net of about $2,000 from the award.

Settlements, or compromise and release agreements are another aspect of the adversary process which remains mostly hidden from view. The National Commission condemns this as a method of claims closure in most instances. It found the supply of meaningful information lean. Like many others, it suspects many misdeeds occur for the workman. For the most part, it appears that a workman is out-the-door after signing a compromise and release agreement.

Although the compromise and release agreement is less costly to the employer, the insurer, and the administrative agency, the workman too often loses both his benefits and his legal rights. Page and O'Brien,[8] reporting on a study of settlements made, find that one state leaves 81 per cent of the claimants surveyed without medical or legal help and therefore, without recourse.

TABLE VI
COMPARISON OF HEARING ORDERS WITH C & E[1] DETERMINATIONS OREGON

Year	Hearing Orders Increasing C & E Determinations		Hearing Orders Increasing PPD Award Issued by C & E		Average Degrees Awarded[2]			Hearing Orders Granting a PPD Award When None Were Issued by C & E		
	Total Orders	% Pro Claimant	No. of Orders	% of Total Orders	By C & E	By Hearing	Average Increase	No. of Orders	% of Total Orders	Hearing Order Average Degree Awarded
1968	708	66.0	308	43.5	38.8	70.5	31.7	99	14.0	29.2
1969	886	73.8	442	49.9	39.8	76.1	36.3	109	12.3	38.0
1970	810	74.9	403	49.8	37.5	83.5	46.0	116	14.3	42.1
1971	955	75.8	478	50.0	43.2	91.2	48.0[2]	115	12.0	44.1

[1] Closing and Evaluation

[2] One degree of disability = $55.00

This is why the cost of compromise release agreements to the individual claimant-workman and to the rest of society is immeasurable. We can also presume that once a workman is shorted of money and his workmen's compensation benefits in this manner, he remains an adversary to the system in the future. These become future costs we haven't seen yet.

Considerations of the costs of the adversary process also extend beyond legal fees and court costs. There are other direct and indirect costs. Certain of these costs have a direct bearing upon the workman involved in litigation. Other costs arise indirectly and are side issues not usually considered.

At the very least, litigation means delay. This can amount to an unnecessary continuation of temporary total disability payments to the workman. This is a direct cost which ultimately leads to increased premium rates in workmen's compensation insurance. It also becomes a cost ultimately passed along to you and me as consumers of products and goods.

The cost of delay is also borne directly by the workman. Many workmen are unemployed during the time their appeal is going through the appeals mechanism. This may cause a workman to seek welfare assistance, which shifts the burden of cost directly upon the general citizenry. Also, prolonged disability reduces a workman's overall earnings. Lesser earnings means a reduction in the workman's tax contribution to the state and nation.

Other delays also occur. Rehabilitation of the workman is often postponed or delayed until the litigation is completed and this increases costs. Many people expert in the rehabilitation testify to the increased difficulties and costs arising out of the late attempts to rehabilitate a workman.

In his discussion and recommendations on rehabilitation of workmen. Schramm[14] notes the harmful effects of litigation by delaying the rehabilitation. In addition, the delays caused by litigation in rehabilitating the workmen, according to Schramm, often cause temporary disabilities to become permanent. Schramm suggests a need for change in the legal fees system to encourage immediate rehabilitation.

Aggravating this delay commonly seen in rehabilitation and recovery is the fact that litigation also brings forth conflicting medical opinions. The workman is subjected to more medical examinations from both sides. Whereupon, the workman is left wondering who is correct. These elements of doubt created by the adversary process are never conducive to either recovery or rehabilitation.

A hidden cost little recognized but related to the adversary process is the practice of forensic medicine by treating physicians. Because of the physician's awareness that a hearing may be in the offing in each workman's compensation claimant, the physician is prone to conduct otherwise unnecessary examinations.

In order to substantiate his opinions before the hearing officers and the courts, the physician orders such things as X-rays and laboratory procedures not otherwise needed. This serves to increase the costs of medical care and workmen's compensation insurance.

Although these costs serve to bolster the overall visible costs of medical care, this sort of practice by physicians remains hidden from the general view of the public and, in a sense, are hidden costs. This is a wrongful practice which tends to grow.

We are all wronged directly and indirectly whenever excessive medical examinations are made. The most direct effect is upon the workman. Excessive examinations may cause him to believe his injury is worse than he believed. More X-rays and more laboratory tests serve to focus anyone's attention upon his physical problems.

The additional harmful aspect of this mode of practice is the ultimate effect upon the physician. The ordering of ancillary examinations not otherwise needed except to look well in the courtroom becomes an easy habit. The physician's brains tend to slip into the head of his pen. It is an easy matter to cover the bases by ordering X-rays, blood tests, electromyographs and so on.

The physician gradually and unwittingly comes to rely more upon such testings and as a result loses the mental acumen separating the physician from the technician. This then becomes a pattern of practice which carries over into the care of all the physician's patients.

So it is that the costs of the adversary process in workmen's compensation are both visible and hidden. Woefully enough, we close our eyes to these costs. In part, we probably ignore the costs because we see the adversary process as getting something for the workman. But, does it?

THE BEGETTER

The adversary process fathers no solutions for the workman. It serves to focus the workman's attentions upon his rights rather than his recovery. It fosters the notion that disability is a commodity. This perpetuates a system in which the principal effort is to purchase from the workman the disability we believe exists.

No arguments exist which can deny the value of the services the claimant's bar provided for the workman in the past. But the adversary process now serves as a deterrent to the development of a workmen's compensation system that is responsible and responsive.

The present posture of the adversary process keeps the camps divided. It blinds otherwise intelligent men to the needs of workmen. It keeps those who could perfect the workmen's compensation laws of their states by working together, at odds.

The adversary system in workmen's compensation is as archaic as the laying-on of poultices is as a treatment for pneumonia. Neither cure the underlying disease. Nor do they reduce the incidence of the respective diseases. Litigation merely treats the symptoms of the wrongs in workmen's compensation. It does not eradicate the wrongs.

If we are to cure the wrongs in workmen's compensation, we must legislate with the needs of the workmen in mind. The emphasis in new laws must not be so centered upon the workman's monetary needs. If these laws are properly framed, the rights of both the workmen and the employer will be kept without the chronic need to litigate.

Litigation serves more the need of those who oppose the concept of workmen's compensation than it does the workman. It offers the opposition the opportunity to muddy the waters and deny effective legislative changes in the state's acts.

For workmen, the visible and invisible costs of litigation are too high. Were it not for these costs, substantial benefit increases could be made without any increase in premium rates.

However, the greatest cost of litigation is in its crippling effect upon the workman. And this does not justify our continued support of "the sacred cow."

Chapter Four

DISABILITY AS AN ENTITY

POTPOURRI

THAT OCCUPATIONAL DISABILITY exists is undeniable. How many workmen are factually disabled by work injuries and illnesses is questionable. However, it is reported that more than two million of us are disabled annually by work-related injuries. And, this is what the stew is all about in workmen's compensation—the disabled workman.

Like the common cold, disability is commonly with us in workmen's compensation. We fuss and we fume and we treat symptoms. Unfortunately, our symptomatic treatment of disability consists primarily of large doses of legislation.

But somehow legislation doesn't make disability go away. It doesn't cure this condition. Indeed, at times, legislation seems to aggravate this entire problem seen as the disabled worker.

That we are failing in our search for a cure of disability is evidenced by the injury facts reported. These statistics reveal important trends in work injuries.[3] While they show a reduction in the number of disabling injuries, they also show an increase in the average number of days the workman is disabled.

Disability is a confusing matter to all. This is readily obvious if we consider the number of textbooks and manuals published on the subject of disability evaluation. Sundry formulas and schemes exist in these publications. Each is written in an effort to provide a method for the equitable evaluation of a workman's disability.

But what happens after we apply our chosen disability formula and arrive at a fair disability rating? We are further confused to find our disability rating bowled over by the pontifications of hearing officers and judges.

Underlying our quandry, in part, is our present-day approach to the injured workman. In general, we treat him and care for him until he is obviously permanently disabled. Whereupon, we find ourselves busily chasing after his disability, trying to flag it down with dollar bills. As a result, we skip over the reasons why a workman is disabled and focus our attention upon the monetary aspects of his disability.

By giving most of our thoughts to the financial aspect of disability, it leaves employers, insurers, state agencies, doctors, and our workmen floundering in a monetary morass.

"Is he disabled, or not?" This is a major question we ask ourselves after a workman reports an injury. If in our eyes he is not, that ends the matter, unless the workman chooses to see an attorney. But, if we see the workman as disabled, we step deeper into the quagmire by immediately asking, "Jeez! How much will this one cost us?"

Our experience with disability is voluminous. Our solutions are short-sighted. Our research is nil. Our reluctance to probe deeper is likely our natural tendency to resist any change. Change requires effort and because of this it may be easier for us to stay in the stew.

THE MILIEU

The present environment in which disability exists is polluted with past thinking. There is a general presumption that injury is the cause of disability. And this is very nearly the depth of our thinking. If someone dares ask, "Why is this workman disabled?" too often the reply is, "Because he is injured."

With this assumption the claim is mechanically ground into the system. Too often it ends much later with the legal minds prevailing. To the legal mind the disability exists in part because of the workman's age, education, and training. They combine these factors with the physical impairment and this becomes *why* and *how much* the workman is disabled. This is merely a method of arriving at some dollar value for a disability. It is not a solution to why a workman is disabled.

Present methods too frequently accomplish a *zero* result for the disabled workman. Our bickerings disable him more. It also limits our own thinking in seeking out the causes underlying any disability. Working with these limitations we then insist upon salving the workman's wounds with the big pay-off, or, we sometimes attempt the arduous task of rehabilitation once the workman is sorely disabled.

As citizens, we are each responsible for the present environment existing in workmen's compensation. We leave too much of our own responsibilities to the medical and legal profession. We see it as their task to find the solutions for the disabled workman. We view our own obligation to the injured workman as one of financing those solutions found by doctors, lawyers, and judges.

For example, we prevail upon the doctor to determine when it is that a workman is ready to return to work. But we give our doctors precious little assistance. We resist creating a work situation wherein a workman can resume working before he is completely healed.

In looking to the legal profession, we often ask them to make a fair determination of a workman's disability. However, little is done prior to their decisions to create a work environment for the workman wherein his age, or education, or type of injury is less disabling.

Doctors, lawyers, and judges are also looked upon as the experts in the world of work. Some of these men are knowledgeable in the work world. Many are not. Few of these professionals ever enter the work place. Thus, they are likely to have a visual image of the work-world that is yesteryear's.

The workman rightfully declared disabled in yesterday's work-world may not be today. Technology is daily changing the labor required in our jobs. Because of this, the labor of work is in a fluid state and this profoundly affects the degree of factual disability in a given workman.

So it is that we place a disproportionate amount of the burden upon the legal and medical professions by our demands that they alone solve the problems of disability and the disabled workman. The rest of us tend to confine our concern to complying with workmen's compensation laws we create, rather than looking to the true needs of the disabled workman.

Our general approach to the injured workman clings to the proposition that injury begets disability. Then we nail down our presumption by shucking our responsibilities to the workman in a lawful manner because compliance to the law is the big thing today in workmen's compensation.

The present chaos in workmen's compensation calls for reform, and it is time. However, to find the solutions to the disabled workman we must begin again at the beginning.

ORDER OUT OF CHAOS

"What is the cause of disability?" This is a mind-numbing question. There are so many causes. Any attempt to document all the causes of disability is a lesson in futility. And if it were done, the volumes filled would be useless.

The causes of disability are individualistic and pertain mainly to the specific person disabled. However, there are three major components common to every instance of disability.

All disability is initiated by some one thing, or things. Once it is initiated, a disability persists only if reinforcing factors are present. Finally, all disability exists within the parameters of time.

Viewed in this manner, it is possible to put all factors causing a specific workman's disability into logical order. Since all causes of disability fall within one of three components comprising disability, this enables us to classify the three components of disability as:

1. Initiators
2. Reinforcers
3. Time.

In the next chapter we will explore the usefulness of this three-component theory as a possible means of predicting permanent disability.

Meanwhile, further discussion of the three components is in order. Learning to recognize those things which initiate and reinforce a person's disability gives us insight into the means of preventing disability.

In preventing disability, just as the parameters of time work against us as a cause of permanent disability, time also affords us a framework within which disability prevention is possible if we act. But, it requires that all parties to a workman's claim act with understanding.

THE INITIATORS

Lawful Initiators

For purity's sake, injury should not be considered as an initiator. However, the nature of our present workmen's compensation laws relates injury to permanent disability as a cause. Therefore, injury is considered as a cause, but as a cause of permanent disability, it is considered so only as a cause in those circumstances specified under our workmen's compensation laws.

An example of injury as an initiator of permanent disability is found in the Oregon Workmen's Compensation Law. This section of the Oregon law reads:

656.214 Permanent partial disability

(1) As used in this section:

    (a) "Loss" includes permanent and complete or partial loss of use.

    (b) "Permanent partial disability" means the loss of either one arm, one hand, one leg, one foot, loss of hearing in one or both ears, loss of one eye, one or more fingers, or any other injury known in surgery to be permanent partial disability.

(2) When permanent partial disability results from an injury, the workman shall receive $70 for each degree stated against such disability as follows:

    (a) For the loss of one arm at or above the elbow joint, 192 degrees, or a proportion thereof for losses less than a complete loss.

    (b) For the loss of one forearm at or above the wrist joint, or the loss of one hand, 150 degrees, or a protion thereof for losses less than a complete loss.

    (c) For the loss of one leg, at or above the knee joint,

150 degrees, or a proportion thereof for losses less
than a complete loss.

(d) For the loss of one foot, 135 degrees, or a proportion
thereof for losses less than a complete loss.

Notice that the language of the law for scheduled injuries is
couched in terms of "permanent partial disability." All state
workmen's compensation laws contain "scheduled injuries" as a
section of their law.

Let's see how it is that injury becomes the cause of disability
under the scheduled injury section of the law: Gordon B. reaches
under the chain of his machine. Gordon is a fast worker. The
machine is still running. The frayed edge of his sleeve is caught-
up between the chain and some very large gears. He screams.
Someone slams a fist down on the stop switch. But before the
machine can stop, Gordon is left flopping about on the floor,
separated completely from his right arm.

At the very moment that last shred of tissue in Gordon's arm
parted, Gordon was permanently partially disabled by the language contained in the law.

Injury is synonymous with disability in this instance—by law.
The law gives no quarter to the fact that Gordon B. is an individual. True, Gordon B. has a serious physical impairment. But
is he disabled?

The law initiates all disability arising out of scheduled injuries. In addition to this, the degree of disability initiated can
be very large as seen in the instance of permanent total disability.

Unappropriated Initiators

Excluding those disabilities initiated by the wording of workmen's compensation law, unappropriated initiators fall into one
of two general classifications:

1. Pre-injury initiators.
2. Post-injury initiators.

The pre-injury initiators are those initiators pre-existing the
injury. Many times they are not ordinarily apparent until after

the injury. Because of this, it is not always possible to prevent the onset of a disability where pre-injury initiators are involved.

However, pre-injury initiators are sometimes recognizable. If recognized, preventative action prior to an injury is possible, as we shall see in the following example:

Larry G. has had several disabling injuries in the plant. None were of long duration. A week on most occasions. One or two injuries kept him off the job two weeks. However, the injuries were rather trivial on all occasions—a strained wrist, a cut finger, a wrenched ankle, and such.

On this particular day, Larry is in the plant's medical office asking the nurse for aspirin. His head is splitting. In talking, Larry grumbles, "Man, that foreman is a gut wrencher."

They talk awhile longer and Larry goes back to work. The nurse, however, has discovered an initiator, the reason why Larry seems to suffer so greatly from so little.

Larry and his foreman tolerate each other and this is the sum of it. Larry likes working for the company, but he's always glad for an opportunity to be away from his foreman whenever possible.

The nurse tells her finding to the personnel officer and he takes the steps that stops Larry's disabling injuries. It doesn't stop Larry from having injuries, but they no longer disable him. The personnel officer transferred Larry to a different job under a new foreman. Larry and his new foreman work well together.

The majority of pre-injury initiators are not this easy to recognize nor to eliminate. But we might anticipate their presence in most workmen because of the commonness of their existence in people.

The ingrained pre-injury initiator commonly exists in most of us. This specific initiator is our own inner desire to be rewarded for personal injury or sickness.

As a pre-existing initiator it is so ingrained in the lot of us it seems almost inherent in the human being. However, it is a learned trait.

Consider for a moment the earliest days of your infancy. Hunger or dirty diapers were likely your first known discomforts. Assuming you had the average parents, your raising a fuss over your predicament brought immediate feedings and diapering. However, along with this feeding or diaper changing came a little something more in the form of a little cooing and coddling.

Petting, pampering, or cuddling served as a reward of sorts during your infancy. While they may not be wrong things to do, they do teach us to expect a reward of one kind or another for our discomforts. This expectation quickly becomes a hidden desire of ours.

Of course these rewarding occurrences are not limited to our infancy. Perhaps you were the venturesome one in your family who flipped out of a tree, breaking your arm. After the doctor casted your arm, when you were safely home again, you sat back and reaped the rewards of your misfortune. While one of your brothers chopped the wood, the other mowed the lawn. Prior to your injury, these menial tasks were your specific chores to perform.

Rewards come in all sizes and shapes—a much needed rest, or an escape from drudgery by flying away to Tahiti. Not uncommon are the daily accounts in the newspapers, or radio and television broadcast, telling the story of the injured man or woman. In a manner, this serves somewhat as rewarding recognition to the injured one.

What does the military do for us? As a Foot Slogger, a Swabjockey, a Flyboy, or a Marine, the least you receive for a battle wound is a Purple Heart. It is a proud decoration designed to tell you that you are deserving of something special for having suffered your battle scar.

Medals and monetary rewards are often combined in civil life for acts of bravery resulting in injury to the hero. Then why does it puzzle us when a workman expects a *reward* for his occupational illness or injury?

This brings us to a conflicting point of view existing in workmen's compensation; a point often bringing wry smiles to the faces of otherwise intelligent people because workmen often speak and think in terms of *reward*. When a workman's claim is concluded, and a permanent disability exists, an award is given. It is considered an award because it is based upon a determination of judgment.

However, to the workman, the award is a more straightforward mental process, as evidenced when he simply says, "My reward." This is an honest statement of thought on the part of a workman because it is backed by a lifetime of rewards ingrained in his mind.

Other pre-injury initiators are as numerous as are human beings. This is so because our minds, or emotions, our very lives determine this pre-existence. But, given thought, they are recognizable. The next example may seem as though we are reaching into the ridiculous, but we really aren't. While this pre-injury initiator is often buried, it certainly is not rare.

Herbert H. twists his ankle just before the coffee break one morning. It smarts and it swells a little. But Herb goes on limping about his job, a little slower perhaps. His work production is impaired a bit, as is he. He's not disabled however, not until the quitting whistle blows and he limps home and through his front doorway.

Herb's disability is initiated the moment his wife sees him limping into the kitchen. It is then, with all the warmth and meaning she can muster, that she says, "Well! You big boob! I see you went and hurt yourself. Now I suppose you cain't go to work!"

These seemingly unendearing words are what initiates Herb's disability. He's gained her undivided attention, and its the first in a long time. This bedraggled, stringy-haired wife is no longer given to paying much heed to Herb's needs.

Her attention is totally consumed with the chores created by five kids in the house. That, and making Herb's income stretch. For her life is a drag and her husband is a drag. But for Herb, she is his girl; and he misses those affectionate pats, and glances of their early days of marriage.

For Herb her momentary, demeaning notice of him is a measure of gratification to him in these lean times. The following morning Herb remains at home, disabled.

Pre-injury herd initiators also exist. These initiators lie in the laps of all as individuals, and as a society of people. They are presumptive in nature and contribute to our thinking that work injuries are the cause of disability. They are subject to change through education and the efforts of disability prevention.

A work injury or illness is a *badge of honor* for a man in need of such. Take the example of a workman in his early or mid-fifties. He is a man who saw no point in schooling after he learned to read a bit and spell after a fashion. Going to school didn't meet his needs. Getting a job and earning an honest dollar did, and so, he never bothered going back to school after the eighth year.

He met and married her and worked hard for her and their kids. And he dreamed the dreams of hitting-it-big. He never did. He never had the time. Somehow they always lived in rented houses, plain and simple, near the mills. Hand-me-downs and secondhand cars with thinly threaded tires are a fact of life. His life is an unending struggle which he barely keeps together by hard work.

And then the injury. And this man with his worn and weary back never works again. He has given his all to the mill and the bosses and this disabling wound from the battle is his wound of honor.

His family understands and there is a certain pride in them of Dad. In their minds, he gave his best for those bosses. Although slimmer times are ahead because they must now live on those compensation checks, they believe it isn't Dad's fault. He'd work again if he could!

Looking backwards in this situation we see the submerged pre-injury initiator present. The injury triggered his total disability. This man isn't dishonest. He just couldn't hack it any longer. He had no emotional or financial resources to draw upon. His injury became an acceptable solution to his own inabilities, to his family, to his neighbors, and to his co-workers. His continuing poverty is now more acceptable because of the work injury, and perhaps even a little less bitter.

A more common example of pre-injury herd initiators is seen in disabling back injuries. In general, our society is *psyched-out* by the back injury problem as it presently exists in workmen's compensation. And it is little wonder.

Looking at the costs reported by just one state is sufficient to see that back injuries soak up sizable sums. Florida's Workmen's Compensation Bureau [6] reports back injuries represent 8 per cent of the reported injuries to the trunk area of the body. Their report also reveals that 1,923 back injuries cost the state of Florida in excess of ten million dollars in 1970. In addition to those high costs, the average claimant lost 683 days from work.

Such figures should tell us something about back injuries if Florida is at all representative as a state. Workmen and their families, employers, insurers, and most of us are running scared of back injuries. As a result, we condition our workmen to over-react to a back injury. And for good reason,

It is no news to a workman looking for work that a history of prior back injury is bad news to an employer. A great many employers are reluctant to hire any workman having such a history. Why else pre-employment physicals?

While employers indicate that they desire knowledge of the workman's general health by pre-employment physicals, considerable emphasis is placed upon determining the condition of a workman's back. Otherwise, why do we do so many pre-employment back X-rays?

The emphasis on pre-employment physicals isn't done so much with the intention of matching a workman's physical condition with the job to prevent further injury as it is to weed out the culls having poor backs.

Workingmen and women know this. Their families also know that a back injury has a far greater impact than the immediate effect of the injury. Even though the workman recovers completely from his back injury, he knows the hex is upon him.

Knowing this may be a factor even in short-term disabilities when a workman injures his back. It is likely many workmen remain off work longer than necessary after straining their backs. They do so to be absolutely certain their backs are healed. They feel their employer is more likely to tolerate an extra week or two of total disability than he is to a second or even third lay-off because of recurrent back pain.

Then there is this matter of back surgery. Walk down any street. Stop a variety of individuals and ask, "Do you believe a back operation is apt to disable a workingman?"

"Yes," is the answer given more commonly than not by the average citizen. In general, each one of us can recite at least one person we know who is still sorely disabled in spite of back surgery. And many believe the surgery is what done-the-poor-fellow-in.

Many physicians are now turning away from certain surgical procedures on back injuries. Many orthopedic surgeons are reluctant to do spinal fusions today. This is not based upon a lack of competency by the surgeons. They too often see this surgical procedure failing to help many workmen and, it may even disable the workman to a great extent.

However, before we condemn back fusions completely, let's consider other factors. Back fusion failures may not be due to the procedure. Here again, back surgery is bad news to employers, insurers, and most others. Herein may lie the culprit.

A workman undergoing a back fusion is facing some odds. He may never return to his old job and his old employer. On the other hand, he may. But, more likely than not, he isn't positive his employer will have him.

In addition, he knows he is facing the fact he may not be helped by the surgery.What if the operation is a failure? Will he ever work again? He knows his vocational future is at stake. He too has heard the many sad tales.

Surgery is frequently done upon a good many people who are not psychologically prepared, and back fusions are one such surgery. Many fears pre-exist back surgery but remain submerged and unrecognized.

For a workman having strong fears, remaining badly disabled is also a means of survival. This man knows in his mind prior to a back injury that if he hurts his back badly, or if he should require surgery, total or nearly total disability is a way of protecting himself and his family to some extent; that they can survive through the support of workmen's compensation, social security, Veteran's Administration, or even welfare, as a last resort.

This pre-injury initiator is preventable in most workmen. We can reduce the number of surgical failures arising out of back surgeries performed under workmen's compensation insurance. It requires that we begin by doing at least two very necessary things preoperatively:

1. Preoperative psychological preparation of all workmen undergoing back surgery.

2. Preoperative vocational planning with reasonable assurance
   that the workman can look forward to a gainful occupation
   after the back surgery.

The post-injury initiators are mostly preventable. But to eliminate post-injury initiators from the scene requires that we study
their nature. Our research efforts here should be as intensive as
they are for cancer, or heart disease, or accident prevention.

The importance of such research is recognized if we consider
some facts. More people suffer from the effects of disability than
from cancer or heart disease. In addition to this, we are not likely
ever to be totally successful in preventing accidents causing
occupational illness and injury.

Wherever accident prevention fails, the potential for a disabling injury is present. But, we still persist in putting our efforts
into accident investigation and no effort into researching the
ways to prevent disability.

While the pre-injury initiators lie within the man and his past,
post-injury initiators lie in our actions and the reactions within
the injured man to our actions. Through our actions we cause
the disabling responses that initiate disability in occupationally
ill or injured workmen. The disability thus begun may be large
or small, long or short, temporary or permanent.

A thing as simple as a few words from a fellow workman may
initiate an injured man's disability. For example: John D. burns
the palm of his hand on hot metal while working. It isn't a large
burn, but it is second degree. He has it dressed in the first-aid
room and goes back to work. It is painful, and John has difficulty
handling his job.

John continues to work, mostly one-handed, until Bill says,
"If it's hurtin' you that much, why in hell don't you go home?
I sure would. They got insurance. Besides, I knew a guy once
that almost lost his arm from a burned hand. Got infected. You
know! All this dirt and stuff got into the burn. Be smart, pal!"

"Maybe I will," John D. replies. Bill pats him on the shoulder,
smiles and says, "That's the way, pal. You're gonna get paid anyway. And don't come back until it's healed good. And see your
Doc!"

John D.'s hand healed well but he didn't come back to work

for more than a week. The doctor said he could, but John D. didn't take any chances.

No permanent disability resulted from this injury. And although the disability was of short duration, John D. learned something about being hurt. Neither his foreman nor anyone else in the plant contacted him or made any effort to keep John D. working. There were other jobs, clean jobs, but they made no effort. Thus, John D. learned the following things about being hurt: leave the job, take a rest, the boss has insurance.

Company policy is not an uncommon post-injury initiator. Many companies still adhere to a policy which amounts to the principle: *if the workman can't work his regular job, he can't work.* They insist upon the workman being well enough to handle his regular work before they permit him to resume working.

Such a policy automatically disables the injured workman when an injury prevents him from working his regular job. He is forcibly disabled by policy rather than by fact.

It is difficult to fathom the reasoning of a company that forcibly disables the injured employee. Some use the excuse that they fear the workman will further injure himself. Others say it cuts production and raises costs. Regardless of their reason, one thing such a policy does is to teach their workmen to use trivial injury as a means of getting extra days off, or a paid vacation of sorts in the name of disability.

Organized labor is also a party to a similar circumstance. Contracts drawn between a specific union and a company may prohibit the company from shifting workmen about in their jobs. When this is the case, the injured workman has no alternative. He must remain on time-loss until he is sufficiently healed to return to his regular job.

Then, of course, there is the instance where a workman is involved in an accident and seemingly uninjured. But because a piece of expensive equipment is damaged, the employer fires the workman.

The employer is then stunned to find the workman filing a back injury claim. To the employer's chagrin, the workman remains sorely disabled for some time, perhaps on a permanent basis. Although this is an act of revenge by the workman, who can see a sore back? And it takes some doing to prove him wrong.

Or take the situation where the employer insists that an injured workman go to a designated physician. The workman may have preferred a different doctor. The injury is slight. Because it is a slight injury, the physician to whom the employer sends the workman rightfully insists that the workman return to work. The workman pushes back because he misunderstands. The workman goes home instead, mumbling about the "Damned Company Doctor!"

A doctor's ignorance of the workman's specific job sometimes initiates disability. For example: a workman slips and sprains his wrist and goes to his doctor. It's the workman's left wrist and the workman is right-handed. The doctor asks, "What do you do?" and the workman says, "I sort Handy-Dandys."

"Are they heavy?"

"Yeah, Doc. About fifty pounds each."

"Well, you won't be sorting Handy-Dandys for a week or so," the doctor tells the workman. What the doctor doesn't know is that Handy-Dandys are sorted by a machine. The workman has about four keys to punch with his right hand. However, the workman believes in his doctor. And, following his doctor's advice, the workman stays home for the next ten days.

Post-injury initiators and their examples are innumerable: a word, a glance, an attitude, a policy, missed communications, anger and so on serve to initiate disability. Perhaps not all are preventable in the true sense because people and their actions and reactions are involved in illnesses and injuries.

The nature of an injury in the beginning does not predetermine the outcome, in terms of disability. Frequently we see simple injuries becoming monstrous claims problems in workmen's compensation.

These perplexing claims problems are what finally goaded the California State Compensation Insurance Fund into a research project through Teknekron, Inc. After a study of 600,000 claims, its findings caused the Fund to implement a new claims handling process called the "Claims Director System." This system is reportedly yielding good results in California because it humanizes the claim.

The California Fund is seeking to eliminate those stressful events Leavitt[9] refers to in his writings. They are turning away from the cold, impersonal claims handling system of the past. Their claims director system brings back an interpersonal relationship between the claims adjuster and the workman. The claims adjuster is no longer a nonentity to the workman.

However, even though an insurer utilizes the best of claims handling systems, it cannot always control the other parties involved in a claim. A claim may well take a sour turn because the workman is stressed by the actions of another party.

A party to every claim which is particularly important to the average workman is his employer. Meineker[12] finds the blue collar worker values security needs, material needs, working conditions, co-workers, and ego needs in this descending order.

Like it or not, there is a parental relationship between the employer and the average worker. The employer who ignores the injured workman is apt to initiate a permanent disability in the workman.

Especially after injury, the workman's relationship with his employer is important. Many large employers do recognize this need of affiliation with a workman. These employers visit their employees during the employee's course of recovery. They keep in contact, knowing full well the positive psychological impact this has on their employee.

In contrast, the small employer appears more likely to stress the workman. The small employer is likely to rely more upon the compensation insurer to carry out functions which should be the role of an employer instead.

Many small employers are factually ignorant of workmen's compensation law and its workings. As a result, many small employers fear reprisal by the workman. Because of this, these employers view workmen's compensation insurance as protection against their injured employee.

With such impressions and attitudes, a small employer is likely to stay clear of the injured employee. If an employer has an underlying fear of talking to the employee about the injury, it is likely that the employer fears his words might be incriminating if the employee sues.

A sense of guilt is also more likely in the small employer than the large employer. The small employer may feel a certain sense of fault for the accident. A person who feels such guilt never likes to face the one he believes he has harmed.

Injury, especially serious injury, normally creates emotional responses, one of which is sympathy. The closer the relationship between an employer and an employee, the greater the response. And this can be a problem to the small employer who is perhaps closer to his employees.

A small employer's past emotional responses may now be the very thing which causes him to avoid an injured employee. Out of sympathy, this employer may have made commitments to an injured employee in the past that could not be kept. It then is easier for an employer with this temperament to avoid seeing the injured employee. It is better to do this, the employer feels, than to face the embarrassment of an unkept promise later.

Finally, the attitude of many small employers is the result of insurance experience ratings. Singularly, many do not qualify for experience ratings that lower their insurance costs.

A small employer boxed-in by experience rating isn't really encouraged to give as much consideration as he might otherwise to either accident prevention or disability prevention. As a small employer, he may feel his premium rates won't change anyway. This may cause such an employer to leave it to his insurance company to function in any manner it chooses.

So be it. Whatever the reason the employer has, he is gambling whenever he ignores an injured employee. While the injured employee may be totally disabled on a temporary basis at the time, there may be no permanent disability in the offing until it suddenly dawns upon the employee his employer doesn't give a damn about his recovery.

This sort of neglect initiates permanent disability in an employee through one, or more, of many ways such as anger and revenge, fear of the future, depression, or perhaps the underlying desire by the workman for his employer's recognition of him; and, a permanent disability is certainly a means of gaining the attention of most employers.

All of this seems to be putting the blame for permanent disability in workmen upon insurers and employers. However, they are not the sole source of post-injury initiators. Post-injury initiators of permanent disability also spring from life around a workman.

Take the friend who says with the best of intentions toward a workman, "Man, you're in trouble. Take it from me, I know! Look at what happened to Walt. Hurt his back working for that outfit. He's had trouble ever since. That's the way it is. Hurting your back at work is bad news. It'll bug you the rest of your life. Take it from me! Make sure they give you some sort of settlement."

This friend means well. At the time, you as the workman may not give his advice more than a fleeting thought. But later, you think about it again. And still later. Maybe it's true. Danged if your back isn't nagging you. Little things that didn't bother your back before seem to now. Besides, the soreness isn't going away like you expected. Maybe Bill's right. What if rheumatism, or arthritis, or something else sets-in? Maybe you'd better see about a settlement.

Many things in a workman's life may initiate a permanent disability in him. However, once a disability is initiated, other factors must reinforce the disability if it is to become permanent in nature.

THE REINFORCERS

Disability demands reinforcement, otherwise it withers. This fact is evidenced in the claim files of men previously declared permanently and totally disabled. Some of these men are living useful and productive lives today. They remain impaired but not disabled.

A disability is the tangible response of a person to his or her predicament. The person remaining disabled does so because disability is the solution to the situation. It is those conditions of a workman's situation which serve to reinforce his disability.

However, disability is not a stationary, self-sustaining matter. Permanent disability is not possible unless it is reinforced in a repetitive, continuous manner. Some reinforcing factors are common to all workmen. Others are not. Most disability is reinforced by combinations of both.

In looking for factors that reinforce a workman's disability, we must keep in mind both the commoness of certain factors in all people and the individuality of the person. What follows in these writings stresses both the commoness of man in relationship to other men and the individuality of man.

Personalities, Plus

Some reinforcing factors are highly personal in nature. They reinforce our particular disability because we are what we are. Our needs are uniquely our own.

The story of Herbert H. used in our discussions of initiators is again useful. His personal needs initiated his disability. His own needs also reinforce Herbert's disability.

In summary, Herbert H. limped into his house and gained the eye of his wife. Her momentary, obtuse notice of him initiated his disability. He didn't leave the house after supper that evening. Nor did he stumble out of bed the next morning, snorting and thumping his way to the bathroom.

There was no banging of lids nor rumblings of the corroding water pipes. The fact is, Herbert didn't get up. Later his wife awakens and, peering sleepily at him through a few stringy strands of graying hair, she asks, "Ain'tcha going to work?"

"Nope," he replies. Whereupon she mumbles, "Figured you weren't. First night since Christmas you ain't gone down to the tavern. It must be paining you more'n most things."

In the ensuing week these sparse, but continuing conversational exchanges between Herbert and his wife keep him confined to hobbling about the house unable to work. Neither he nor his wife recognize his needs and the true nature of his disability. Both look to his sprained ankle as the cause of the disability.

A slightly different twist to the personal needs of the individual workman comes out of Beals and Hickman's study.[10] Their study shows that the man who pictures himself as the ultimate in rugged individualists needs disability.

This type of workman needs to be either a complete, whole man able to excel in his work, or he needs a disability that completely prevents his working. This workman cannot tolerate being in any other role than that of the biggest and the best man on the job. Therefore, impairment totally disables this workman.

The Person and the People

That the type of bodily injury you suffer as a workman has a profound effect upon you and others around you is a fact. Two common types of injuries exist in workmen's compensation that are good examples. They are: (1) major amputations and (2) back injuries.

Perhaps you already know two such people who have had these injuries. If so, stop and think for a moment. What is your attitude toward the amputee? Then, what is your thinking about the one with the bad back?

Let's explore for a bit something which occurs with the guy who has a back problem. Can you see his bad back? Maybe so if it is bad enough. Perhaps the way he walks catches your eye, or the manner in which he rises from a chair, or maybe the way he lifts or avoids lifting objects. But what you think you see as a man afflicted by a very bad back may not be true. This man's back may not be as bad as it seems. What you may be seeing is a man protecting himself against the world.

Presume for a moment that you have a bad back. On one or more occasions you have injured it while working. You finally learn after each recovery that it will happen to you again if you don't protect yourself. How then do you protect your back?

Most men and women protect themselves from repeated back injury by words and actions. Being human, they're not likely to wear a sign proclaiming, "Bad Back!"

Their course of action is then limited to more accepted methods of notifying their friends, fellow employees, and maybe even their employer, that they can't perform certain tasks. So, if you are similar to most people, you do the same.

You let others know about your bad back by talking about it.

You gradually assume certain stances and gaits to impress upon them that you do have a poor back. And, of course, you become increasingly aware of your back yourself. You think "back" and you act "back" and in time your life style is the "bad back." Gradually you find yourself increasingly disabled because of your back, and this is a reinforced disability.

What of the amputee? The fellow with the arm or the leg, or both, gone. His impairment is obvious to other people. As a result, they expect less from him. Very likely, the standards of performance most of us set for the amputee are generally less than those we set for a man whose impairment is not so visible.

The amputee doesn't need to assume the posture of the disabled. If the amputee performs a normal or near normal function, it appears to many that the amputee excels in life. Perhaps this is why, in many minds, the amputee is often seen as a fellow who exceeds.

Because an amputee's impairment is so visible to most of us, there is a tendency for us to encourage amputees. We cheer them on to partake in life. This diminishes rather than reinforces an amputee's disability.

Amputees, in turn, urge each other to greater physical achievements. Collectively, amputees have an esprit de corps which cries, "We may be impaired, but we sure as hell aren't disabled!"

But who cheers on the man or woman with the bad back? Somehow their impairment raises questions of doubt in most minds. Because we don't see their pain, or the damaged ligaments, or the injured bones and cartilages in their backs, we tend to be skeptical. In which case their words and their actions tend to worsen our own attitude towards the man or woman who complains of a poor back.

Picture yourself in this situation of having a chronic back problem. What pride do you feel in excelling? Who around you recognizes your achievements?

Taking it all into consideration, it seems that a workman with a bad back functions daily in the atmosphere of the loser. This may account for what we see happening whenever a group of workmen having back injuries and back complaints, associate together in a hospital or rehabilitation center.

This type of association in a rehabilitation center where most of the attendees have bad backs seems to create a situation wherein the born-loser syndrome blossoms in these workmen. Collectively they begin the wail, "We're disabled!" This produces an atmosphere in which they collectively reinforce each other's disability.

The People and the Person

Our workmen's compensation laws are a reflection of our social thinking. Specific areas of these laws as they are now written serve to perpetuate and intensify the workman's disability. Thus it is that societal attitudes come to directly reinforce the disability.

The schedule of injuries as written into our workmen's compensation law declares the workman disabled by the specific type of injury suffered. The workman is given a predetermined "disability award" for his disability resulting from a scheduled injury. This serves both to initiate and to reinforce the workman's disability. Such law gives no opportunity to evaluate the individual as an individual.

Any law implanting and perpetuating those things in a man's mind which are harmful to him is wrong. It is agreed that a workman experiencing any injury classified as a scheduled injury is impaired. However, the presence of a disability is dependent upon the individual and the society in which he lives.

The harmful aspects of predetermining disability which we believe arises out of specific types of injuries such as those seen in scheduled injuries extends far beyond the workman. The law also implants the *disability thought* in the minds of many others such as employers, insurers, workmen's compensation agencies, attorneys, and some physicians. This serves to inhibit the thinking of all involved.

Oregon's workmen's compensation law serves as the example of predetermined disability that is scheduled into law. In the discussion of initiators we saw the schedule of permanent partial disabilities as written into Oregon's law.

Oregon also specifies certain injuries in its law that automatically make the injured workman permanently and totally disabled. This section of the law reads:

"Permanent total disability" means the loss, including pre-existing disability, of both feet or hands, or one foot and one hand, total loss of eyesight or such paralysis or other condition permanently incapacitating the workman from regularly performing any work at a gainful and suitable occupation.

The laws of other states also contain similar sections pertaining to permanent partial and permanent total disability. This reflects the thinking of the people in these states.

They predetermine what it is that makes an injured person permanently disabled. They create a schedule of fact which then denies a workman any consideration of his individuality.

What better method exists than a schedule of injuries written into our laws to both initiate and reinforce a workman's disability?

The Retirement Complex

We are a nation of people with a common goal—retirement. Much of our thinking and time is given to this matter of retirement. We belong to a society which apparently looks upon work as something to retreat from, as evidenced by our national goal to retire our people at an increasingly earlier age under our social security system, and sundry other private schemes.

The plum of retirement is dangled before us. Retirement is a common subject of discussion among workmen during coffee breaks. Work, it appears, is a privilege in the minds of too few.

The vocations of too many people appear to be nothing more than a dreary task. Increasingly, many more workers seem to look upon employment merely as a means of subsistence. It seems that work is unrewarding for far too many, except for the dollars earned.

We see increasing numbers of persons looking to their avocations both for fulfillment, and as a means of escape from daily drudgery. Also, we see workers who merely vegetate and wait, dreaming the dreams of that golden day of retirement. Disabling injury offers then, to those with this attitude, an escape from work.

A disability arising out of an occupational injury can offer an honorable, acceptable escape to a workman weary of his dreary work life. The escape is often only temporary, but it may also be permanent. We've seen how friends and family view the cause of the disability as a wound of honor. Give a man in his late fifties, who is frustrated or even bored with his work life, an honorable means of escape and it is little wonder that he remains sorely disabled, reaching out to pluck the fruits of retirement.

The reinforcement of disability by this type of mechanism is akin to the man who is a failure. As mentioned previously in discussing initiators, the man who fails has pre-existing initiators. So does the man who grasps at early retirement. In both, the disability is maintained by the honorable reinforcement of a permanent disability arising out of occupational illness or injury.

Reinforcement of disability is not the intention of society. Nor is it the intention of the parties to workmen's compensation to reinforce a workman's disability. But we do.

The Other Three

The parties most directly involved in a workman's claim are the physician, the insurer, and the state agency responsible for administration of the workmen's compensation law. In this triad of parties, the treating physician is traditionally looked upon as the pivotal point.

Much of the outcome under today's approach in workmen's compensation seems to rely upon the physician's decision. Some common errors lie in this approach that reinforce disability in the workman.

Allowing the workman a free choice of physicians may be a mistake. The legislatures of many states have been pressured into writing this free choice concept into their statutes, the idea behind this concept being that it is our right as citizens to choose our own physician as we do in our private lives. However, the problem lies in the expertise of the physician chosen in this manner by the workman.

Not all physicians are expert in the treatment of occupational illnesses and injuries. Nor are they knowledgeable in the workings of workmen's compensation law. This is not to disparage medical practitioners. It is merely a simple fact we all must recognize. To thrust an injured workman upon a physician in this situation is a disservice to both the physician and the workman.

A physician, ignorant of the workmen's compensation law of the state, may unwittingly reinforce the workman's disability. This probably happens not infrequently in the matter of claim closure. For example, the physician may have no comprehension of the workman's rights after claim closure, even though the law of that state protects the workman by aggravation rights.

As a result of his ignorance of the law, a physician may extend the aftercare period of the workman. Thinking it to be in the best interest of his workman-patient, this physician prolongs the disability by not declaring the workman medically stationary. This not only lengthens the temporary disability, it also affixes the disability in the workman's mind, and this is a reinforcer that can result in permanent disability.

If a treating physician is not oriented in some degree toward jobs and industries, he unwittingly commits a common error in advice to the workman because physicians think in terms of time segments in tissue healing.

Physicians usually refer to healing periods in terms of two weeks, or two months, or longer. As a result, physicians are prone to release the workman for work on a Monday, or "the first of the month." This prolongs the disability period. It increases the risk of a disability becoming permanent. And it is a definite factor which distorts disability statistics as well as upping the costs of disabling injuries.

Iatrogenic disability is another result of errors committed by the physician. Iatrogenesis literally means the worsening of a condition by the activities of the physician. It is a matter which the average physician is aware of and guards against. Still, iatrogenesis is a constant threat and can occur unwittingly at times on the part of a physician until someone points it out.

The competent, reputable physician never knowingly creates iatrogenic disability. Why then does it occur? Again, it can be viewed as resulting from the physician's ignorance. However, in making such a criticism, we must remember that all men are ignorant, but in different areas.

Iatrogenic disability is sometimes a result of improper treatment by the physician. An example is the physician who is unqualified to set a broken wrist. He may botch the job. As a result of this inept treatment, prolonged treatment may occur, during which the physician hopes to correct his error. Thus a permanent disability is initiated by a poor beginning and it is reinforced by persistently poor treatment, if the results leave a significant impairment of the workman's wrist.

A physician who refuses to refer the injured workman to the proper consultant may well increase the disability by iatrogenic reinforcement. This physician with such a temperament may also look upon all other parties responsible for the care of the injured workman as intruders. In stubbornly refusing to refer the workman he is responsible for unnecessary disability by iatrogenic reinforcement.

A flagrant example of permanent total disability and iatrogenesis exists in the claim files of workmen's compensation. In part it is the blame of an incompetent physician specialized in orthopedics and licensed to practice in that state.

The story of this workman begins with a slip and a fall to the floor. Considerable disability followed in which preexisting initiators had a role. The workman had a preceding history of two unsatisfactory marriages and mental illness severe enough to require hospitalization.

The physician performed an ulnar nerve transplant in the left elbow first, and then an anterior cervical fusion of C5-6 as treatment. The claim was finally closed with a disability award of 20 per cent permanent partial disability of the left forearm and 35 per cent permanent partial disability for unscheduled injury to the spine.

One year after claim closure the workman suffered a second occupational injury in an automobile accident. Treatment by the same physician began with hospitalization and rest, followed by a long series of injections into the neck and low back, and physical therapy. After more than a year of this type of treatment, the workman suddenly developed "ulnar nerve palsy of the right forearm and hand."

An ulnar nerve transplant and medial epicondylectomy in the

right elbow were performed, although electromyographic and nerve conduction studies did not support the diagnosis.

In addition, during the year's interval between this second injury and the subsequent nerve surgery, two other consultants examined the workman at the request of the insurer. Both doctors were highly competent men in their fields of neurosurgery and orthopedics. Both stressed the psychogenic nature of the workman's illness and recommended appropriate psychiatric care.

However, the treating physician chose to ignore their recommendations as well as other recommendations that the workman be referred to a specific rehabilitation center.

Five months later the treating physician performed a two-level laminectomy and fusion on the workman's low back.

For nearly a year after the ulnar nerve surgery the treating physician reported remarkable improvement in the right hand and arm. Then suddenly the workman's right hand and arm required casting because of "paralysis of the fingers."

More surgery on the right forearm and wrist followed the casting. Adhesions of the nerves were severed and a tendon in the wrist cut. The workman's condition was found to be further complicated when the treating physician reported "some evidence of pseudoarthrosis of L4-5." In other words, the prior fusion performed on the low back was not successful.

With some effort, the reluctance of the treating physician to refer the workman to a specific rehabilitation center was finally overcome. At this center the workman had a very thorough examination, including a complete psychological evaluation and an examination of his back by three competent orthopedists.

The conclusions reached by the physicians in the rehabilitation center were that the workman was seriously mentally ill, suffering from schizophrenia. A pseudoarthrosis, or failure of the low back fusion to fuse, was also found. All the physicians recommended proper psychiatric care. All recommended against any further surgery.

Two months after the recommendations of the rehabilitation center were made, the treating physician scheduled the workman for low back surgery.

This is iatrogenesis. Perhaps the disability was not initiated by the physician, but he is responsible for iatrogenic reinforcement which produced a severe disability in this workman.

An insurer is even more likely than the physician to commit errors that reinforce a workman's disability. The insurer has greater exposure to the possibility of error because the insurer has less knowledge of the workman as a person. Until it becomes quite obvious that a particular claim is becoming a monstrous problem, the average insurer makes no attempt to intervene or to make personal contact with the workman, if then.

Meanwhile, the workman's disability is well established and strongly supported by the actions and inactions of the insurer. Leavitt's study clearly shows the importance of the claims adjuster, or claims handler's attitudes and actions. However, the claims adjuster is a reflection of management's attitude.

If management treats claims processing akin to assembly line production, then the claim is dehumanized. The aim of the claims adjuster becomes production. In production a certain amount of spoilage is expected and allowed as a cost of manufacturing. For the insurer, these are claim losses.

Contained in the claims files of a workmen's compensation system is the following claim of a workman showing the effects and results of production-line claims processing:

This is the story of an injured workman and the events leading to permanent total disability four years later. It began when this man fell from a loaded logging truck. As a result, he received a fractured skull with bleeding from the right ear, a fractured collarbone, several rib fractures, and a back injury not specifically diagnosed.

After release from the hospital and while he recuperated at home, this man was referred to an orthopedic specialist several miles away.

Subsequently, the workman submitted his travel expenses to the insurer. Without explanation, the insurer returned a pre-printed form denying payment. Later, however, the insurer corrected its error by allowing expenses equivalent to bus fare, but not at private auto rates.

Meanwhile, the man submitted drug bills. The insurer responded by insisting they have a copy of the prescriptions written. The workman complied. One month later this man asked again about his drug bills. The insurer replied that they were giving consideration to the bills.

Five months from the date of injury, the insurer requested permission from the state workmen's compensation agency to close the claim. The agency denied the request. The man was still under treatment for the injury to his right ear which required chronic drainage by a special tube placed in the ear.

One month later the insurer referred this man to a special examination center operated by the state workmen's compensation agency. The insurer wished a comprehensive examination, hoping this would permit them to close the claim.

The examinations at this center included a full psychological examination and a back examination by three orthopedic specialists. According to the psychologist, this workman suffered from moderate anxiety with neurotic depression. He also had an obvious reading disability as well as a poor intellect.

The prognosis given by the psychologist for return to work was fair to good. The orthopedists found he had a chronic low back strain.

A vocational rehabilitation referral at this time indicated this man was previously a hard working man. He had driven trucks for 30 years. They believed he should be trained to be a truck dispatcher.

Upon the workman's return home, the insurer then referred this man to another orthopedist for a closing examination. One month later the state agency closed the claim, allowing a disability award of 15 per cent to the workman, unscheduled.

In terms of dollars, this award amounted to $2,640. This man's award payments amounted to $182.12 per month. Out of the first check, the insurer subtracted $76.72 because of an overpayment in prior temporary total disability compensation.

Of course, all communications during this entire period were in the form of brief, pre-printed notices, or impersonal short letters. No evidence exists indicating that this workman was ever contacted by the insurer on a person-to-person basis.

The workman immediately wrote a pitiful letter to the insurer the moment he received his first award check. It is obvious from this letter that he was desperate. He explained that he could not pay his bills and sorely needed the $76.72 that was withheld.

The insurer politely replied that they could not pay him the $76.72 because it was the law. However, he could request a lump sum settlement up to 50 per cent of his total award. Or he could request a hearing asking for a greater award.

This man promptly replied, asking for the lump sum money to help pay his bills. However, one month later the workman notified the insurer he had aggravated his back because he had attempted to drive a log truck.

After driving the log truck three days, this man had to quit work. He had gone to a physician for care and paid the physician $39.00. The insurer reduced the bill to $36.90 and refunded to him a check in this amount.

One month after the workman requested reopening of his claim. The insurer sent the following reply, which is quoted in part, "Your telephone request for a reopening of your claim for time loss compensation and medical treatment has been referred to this desk. . . ."

Now our man finds he is communicating with a desk. However, the desk did reopen the man's claim.

That same month the compensation check was waylaid in the mails and the workman wrote the insurer. The insurer sent him more forms to complete and two weeks later he received a second check.

At this point the workman sought care from his family doctor. The doctor was slow in reporting to the insurer and the fellow was again in difficulty.

The insurer received a letter from him because he needed money to buy prescribed medications. He enclosed the drug bills. The insurer denied them because the doctor had not mailed a written report.

One month later the insurer refunded this man for his drug bills. However, this same process was repeated by the insurer two months later.

Meanwhile, a vocational rehabilitation counselor saw this man several times. He referred the workman to two or more potential employers who did not hire him. The workman, meanwhile, found a job in an adjacent state driving a log truck. This job lasted two days until he quit because of back pain.

Following this workman's last effort to work, the family physician wrote a scorching letter to the insurer. The physician indicated that he had known this man twenty years. He depicted the man as totally honest, frugal, and a proud family man.

The physician also claimed the man's present disability was due to fear, frustration, and anger. The physician believed that this man had been in dire need of psychiatric care from the early beginnings of his injury.

Two months later the family physician wrote the insurer informing him that the man was in a veteran's hospital because of a bleeding ulcer. This man had undergone three prior surgeries for an ulcer several years ago. He was operated on again.

One month after this the insurer referred the workman to another orthopedist. This doctor believed the man was unable to work and in need of psychiatric care.

Thus, six months from the time the family physician recommended psychiatric care for the man, he was finally seen by a psychiatrist. In his examination, the psychiatrist discovered this man was suffering from organic brain damage as a result of his injury.

The psychiatrist treated him for five months. The man improved, according to the psychiatrist, and was highly motivated to work.

The family physician also reported marked improvement and a weight gain from 126 pounds to 150 pounds, indicating improvement in mental attitude. Prior to the time he first sought help from the family physician, the workman was so despondent he considered suicide.

Four years from the date of injury, this man was awarded permanent total disability upon the advice of an orthopedist and the family physician. This award lowered his present monthly payments of $200 per month, paid on the basis of the previous permanent partial disability award, to $185 per month.

An interesting last comment found in a document in this man's claim shows this man is divorced as of three months ago.

Many glaring errors exist in this workman's claim. Medical errors, insurer's errors and agency complacency. As a result, this man's work-life disintegrated and his family life was destroyed. The system initiated and reinforced this man's permanent total disability.

TIME

Time is the third component summating with the initiators and the reinforcers that determines the resulting disability. This factor of time is extremely significant in the whole process of occupational illness and injuries. It is within the framework of time that the events causing permanent disability occur.

The effect of time on a seriously injured workman is threatening. Imagine yourself in the situation of a workman suffering the major trauma of multiple fractures resulting from a long fall. You are likely to run the gambit of emotional responses before you are healed.

Your first reaction to the injury is fear. This fear response is usually both immediate and sustaining in nature as you experience the fear of dying, the fear of pain, the fear of crippling and disfigurement, the fear of losing your future.

Imagine yourself being wheeled through a doorway into an emergency room. You are instantly thrust into the medical world and isolated from normal life.

In addition to the fears already filling your mind, these foreign surroundings seem cold and threatening in themselves. Strangely garbed people move quickly about the room, speaking a language which to your ears is the equivalent to Gibbonese.

Filled with pain you wait, you wonder, and you hope as you drift back and forth between the conscious world and that frightening dark void filled with distant voices. And then the pain eases.

You become vaguely aware of traveling along corridors, in an elevator, and around a corner or two. More pain and the fear of more pain fills you. Then you rouse to an eerie sense of stillness surrounding you as you realize you're in a hospital bed.

Through the haze of the following days, perhaps many days, there come two, three trips to a surgery room. You've lost count.

The pain, the pain is always there, dull and deep, and you constantly fear that the pain will sharpen. Those first days in the hospital are enough to tax the nerves of any man.

Later, as the healing process proceeds and the pain eases, you experience conflicting feelings. Your hope mounts and you grow anxious to leave the hospital, but there is also a subtle underlying sense of security and comfort in the routine of hospital life. The thoughts of leaving are also a little frightening.

At the best, hospital life is an abnormal life style, an artificial existence. It isolates the person from family and home, from friends and neighbors, and from the workplace and daily habits.

If, perchance, the emotional sinew of a workman is thin, it is little wonder that the man comes unstrung. In such a man, an untimely injury is likely to crush out whatever emotional resources remain.

Fears and the effects of isolation are not limited to the hospital setting in workmen having severe injuries. These mechanisms continue on through the recuperative period at home. Nor is fear and isolation limited to those workmen having severe injuries.

Any injury confining a workman to home isolates him and brings with it certain fears. Here again the workman is isolated from his job, fellow workers, and life as usual.

The fear mechanism is not as obvious in the workman in this circumstance. It is more difficult to detect, and in the beginning it is likely submerged. However, given time, the isolating effects of the injury potentiates the workman's fears.

A common disabling injury, frequently beginning as what appears as a trivial or minor injury, is the low back strain. Then to the puzzlement of all, it leads to prolonged temporary total disability and perhaps an astounding permanent total disability.

The story of the low back strain is common in the files of workmen's compensation systems. Usually the workman is middle-aged and is involved in a minor accident: a slip, a slight fall, or the wrong lift. This commonly results in the workman remaining home for a few days rest. But the few days rest too often become a worrisome claims problem.

While resting at home, the workman becomes aware of some facts of life: his age, dropping out from high school, years of menial work and limited, if any, future financial security. There is time for worry, doubts, fears and tension.

With tension comes increased discomfort. Tension in anyone of us causes muscle tightness and prolonged muscle tightness causes pain. So it is that this workman then finds his back pain hanging-on. This is an added worry-doubt-fear- tension mechanism that cycles.

The time lengthens and the isolation persists to reinforce the workman's increasing doubts of recovery and work. Next to follow, when the workman's only contact with the workmen's compensation system is the bimonthly check, is an acceleration of the isolation-fear cycle. The workman's disability mounts and periods of idleness increase.

Subtly, his idleness compounds the disabled workman's problems. Given time, idleness is a disabling factor in itself. Forced or assumed inactivity results in a range of emotions: boredom, apathy, and vegetation; all are immobilizing and disabling. Thus, the isolation-fear cycle now comes to include idleness and its disabling effects.

Time and idleness destroy certain habits in all of us, one of which is the habit of doing. Take the work habit, for instance. This is a habit pattern common to most. Going to the workplace daily is usually a basic impetus for our arising. Because it is a habit pattern we follow, it is not any particular problem for most. However, such a routine habit pattern can be broken by will or force.

In this case, the disabling occupational illness or injury becomes the force breaking the work habit. Given sufficient time, the habit pattern becomes that of not going to work.

As individuals, we cling to habit patterns and we resist any change until it is forced upon us. Once we are accustomed to not going to work, we tend to persist in those reasons which permit us to continue this habit pattern.

How quickly our work habit pattern is lost is highly variable in each individual. In part it depends upon such things as a liking for your work and your personal or family needs. However, given sufficient time, the work habit fades in all men or women.

A disability extending over a lengthy period of time also causes a shift in the responsibilities of a workman. With time a shift occurs in the emphasis of the disabled workman's role as a family member and as a citizen.

The workman's responsibility to provide the needed family income changes. If more income is required, it must come by looking to the spouse or other family members, the insurer, the courts, or other segments of society such as a welfare agency. Given time and the right circumstances, we all shuck our responsibilities. In some the sense of responsibility never returns.

Another common human response likely to occur in the workman as the time lost from a disabling illness or injury increases is *transference*. The workman transfers the blame for his injury to a fellow workman, his employer, or perhaps to society in general. This transference of fault may be sudden or gradual. In part this is dependent perhaps upon the threat to the workman's security. However, when transference occurs, the workman's natural reaction is to seek retribution from those he now believes faulted him.

Thus, time is a devastating factor in the disabled. Time allows for isolation, doubts and self-doubts, fear, loss of the work habit, a shift of responsibility, and the transfer of fault.

Man is adaptable and because of this he survives in time. One of man's adaptation mechanisms is the ability to compensate. Compensation to his plight is that which enables a workman to accept a lowered standard of life, and a lower level of activity. In time a gradual realization sets in that time itself no longer has to be met.

Because he compensates, the workman's actions become more those of responses to his whims rather than to his needs. The desire to overcome his plight and his disability wears away. This workman now is ready to accept permanent total disability as his lot.

Time allows for the errors initiating and reinforcing disability after the injury. As an example, let's look at the findings in the claim presently in the files of a large self-insured company:

The claimant is a woman employee. The claim begins because she strained her back. Under the workmen's compensation law of the state, she has a free choice of physicians. She chooses a chiropractor.

Eight months later the claims department of this company discovered this woman on temporary total disability and the chiropractor's bills mounting.

Between her date of injury and the sudden awakening by the insurer that this woman was disabled, the woman quit her job. She had requested that her employer transfer her from the job on which she worked at the time of injury to a job as a secretary.

Although this woman had the capabilities of a secretary, the company personnel department denied her request. Whereupon, she went home. And, she stayed home, disabled for the next eight months.

Eight months is a long time for a back strain, if that is this woman's problem. She did continue working after her injury until her company denied her a job change. This disabled her. Poor claims handling then reinforced her disability. The chiropractor has likely reinforced her disability by overtreatment. These disabling events all came about in a matter of time.

SUMMARY

Disability is the individual's personal response to the summation of the emotional, economic, and social factors arising out of an occupational injury or illness. It has a beginning, a middle, and an end. How it ends for a workman depends upon whether we wish to merely treat it after the fact or come to grips with it in terms of disability prevention.

Chapter Five

PREDICTING DISABILITY

O UR PRESENT ABILITY to predict the outcome when a workman files a claim is nearly nonexistent in workmen's compensation. It is continually amazing the turn a seemingly insignificant injury reported by a workman can take in a disconcerting number of claims. Too often the disability a workman ultimately manifests is entirely disproportionate to his residual impairment. These events leave employers and insurers, as well as the treating physicians and others, flabbergasted.

How is it possible to predict early in the course of an injury and a claim the disability likely to follow? The blame for our present day inability to predict the degree of disability is due in part to the limited research done to date in workmen's compensation. Currently, the luck of the claim is too dependent upon the luck of events following the injury.

The purpose of this chapter is to explore disabling injuries and to present a possible approach to predicting the degree of disability. However, considerable study is still needed to perfect any method proposed at this time.

Part of what makes it difficult to foresee the impending disability in a claim is the claim form itself. The claim form of today is not very different from the claim form of yesteryear. The information gathered on current claims forms appears to have no predictive value for a claims processor.

Research may possibly change the entire format of the claim forms in the future, whereby they can become a predictive tool for the claims processor. Claim forms of the future may not contain any of the data presently considered so sacrosanct on today's form. Again, this speculation remains for the future.

The present task is to work with what is available in the present system of claims management, and to effectively manage a claim in a manner that assists a workman in recovery.

DISABLING INJURIES

Another Way of Looking

Disabling injury is looked upon as a disabling injury in workmen's compensation. It usually ends there. However, there is another way of looking at such injuries. This other way also offers a simple means of classification having practical application in everyday usage.

All disabling injuries are classifiable into one of the following three classifications:

1. The intrinsic disabling injury
2. The conceptual disabling injury
3. The combining disabling injury

Classifying disabling injuries in this manner gets away from the age-old habit of looking at a specific injury as serious or nonserious. It enables us to see why a nonserious injury sometimes turns into a claims monster wherein the workman becomes sorely disabled.

Also, classifying each disabling injury this way enables claims people to evaluate the possible turns a claim may take. It gets them away from claims management by retrospect. It gives each some potential to predict and prevent.

The Intrinsic Disabling Injury

In its purest form, the intrinsic disabling injury is by its nature, immediately permanently disabling to a workman. No external factors enter into the disability resulting from this injury because workmen's compensation law declares this workman as permanently disabled.

The disability may be permanent partial or it may be permanent total disability. In general, most intrinsic disabling injuries are those seen as scheduled injuries under workmen's compensation law.

Many examples exist. One example is that of a driver-training instructor involved in an auto accident and totally blinded by his injury. Under workmen's compensation law, this man is considered permanently, totally disabled because he is legally blind.

The workman who suffers a major amputation such as the loss

of an arm also has an intrinsic disabling injury, and so does the workman losing a finger. The degree of disability differs in each instance and is determined to a great extent by the wording of the specific law.

The degree of disability, then, is a matter of legislative edict. All parties to the claim and all involved in the care of the specific workman can anticipate the result. However, the factual disability need not measure up to the legal disability in every instance.

Because factual disability differs from disability prescribed by law, it is possible to measurably alter the result of an intrinsic disabling injury.

For instance, the driver-training instructor need not spend the remainder of his life in the abyss of the unemployed. Factual total disability can be prevented even in this workman. Through properly timed disability prevention efforts, the instructor can be returned to a useful life and a gainful occupation.

On the other hand, if no effort of disability prevention is made, the result can be unreasonable as well as totally disabling. Left only to his own devices and the legal and timely payment of compensation checks, the driver-training instructor will unwittingly accrue unnecessary disabling disability.

The Conceptual Disabling Injury

The conceptual disabling injury, in its purest form, is entirely the result of extrinsic factors. This is the sort of disabling injury we see arising out of the most trivial of injuries, or the almost no-injury accident. However, there may also be significant injury involved.

The disabling effects of conceptual disabling injuries stem from the individual's response to the situation. Emotional instability, precarious economic circumstances, and societal attitudes are the etiological sources of the conceptual disabling injury.

Pre-existing initiators play a large role in such an injury. The example used in chapter four, concerning the workman disabled after spraining his ankle, explains the role of pre-existing initiators in this class of disabling injury.

The length of disability resulting from a conceptual disabling injury is variable. It depends upon the circumstances. The workman wishing to escape from his work for a day or two may use a trivial injury for this purpose.

Permanent total disability is a possibility in another workman having a conceptual disabling injury. Take, for instance, a mentally retarded farm worker. Should he accidentally injure his back severely, the potential of permanent total disability quickly becomes factual. Such a workman is untrainable to other work and now unemployable because his back injury precludes his working as a farm laborer. The pre-existing initiator, of course, is this workman's mental retardation.

Public reaction to a horribly scarred face is another example of a pre-existing initiator in the conceptual disabling injury. Take the example of a store clerk whose face is badly disfigured by a burn injury. There may be no impairment, but significant disability.

The store clerk's disability is brought about by the public's reaction to an unsightly face. Thus, an employer refuses to hire this workman because the employer is afraid the workman's appearance may repulse the customers.

Because of the factors involved, the pure conceptual disabling injury is often the most difficult to recognize. These are injuries that constantly amaze all when they suddenly realize a claim is turning into a disaster. However, workmen disabled in this manner are not malingerers. The malingerer is a fraud. The workman manifesting a conceptual disabling injury is a victim. Unfortunately, many workmen have conceptual disabling injuries that go unrecognized and they subsequently become sorely disabled.

The Combining Disabling Injury

The combining disabling injury is the commonest disabling injury. Few conceptual disabling injuries are purely conceptual in their origin. And fewer intrinsic disabling injuries are purely intrinsic in the totality of the disability.

Most injuries disable the workman through the combining effects of the intrinsic nature of the injury and the conceptual factors present in the situation.

The conceptual factors coming to play are the initiators, the reinforcers, and time. The conceptual factors also have the greatest influence upon the claim in the majority of instances.

However, if either the conceptual factor or the intrinsic factor is great in an injury, the workman is greatly disabled. Between these two extremes is the bulk of the injuries wherein the degree of disability is determined by an intermix of the two factors. This is why the care of the injured workman is not purely the prerogative of the medical profession.

We can get a visual image of this entire concept of disabling injury by visualizing it as a cone of light passing through a pinhole in a card.

As seen here (Figure 1), the outer rays of the cone of light represent respectively the intrinsic and the conceptual disabling injuries. Lying between these two boundaries is the whole intermix of combining disabling injuries.

The light passing through the pinhole and falling upon a surface at any given level represents the various percentages of disability resulting, including permanent total disability.

Heretofore, claims people have classified disabling injuries in terms of the bodily part and the nature of the injury to the part. While classifying injuries in this manner is useful in such research as injury statistics and accident prevention, it doesn't lend itself to claims management.

We know the injury in a claim. For effective claims management, we must know the person, if we wish to prevent disability. The problem in knowing the person whose claim is before the claims worker lies in the general lack of personal information about the person.

Putting a claim immediately into one of the three classes of disabling injuries presented here enables the claim worker to anticipate future events that might possibly occur in the claim.

It brings to the person managing the claim a greater awareness of the person making a claim. If necessary, the claims worker can seek additional information pertinent to the situation. Awareness and anticipation of events, and possibly future events, in a claim is a requirement in preventing unnecessary disability in workmen.

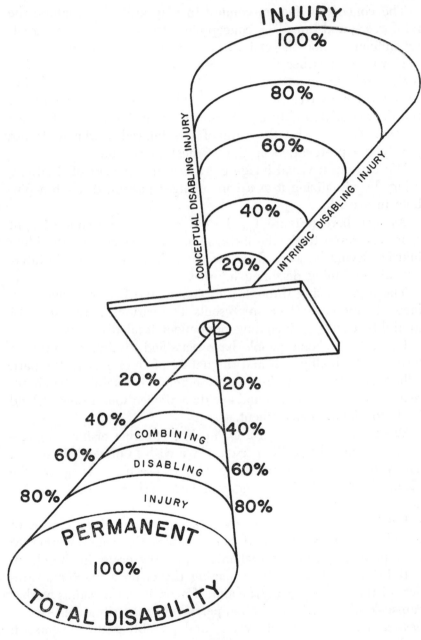

Figure 1.

A PREDICTIVE THEORY

While the preceding discussion is useful in claims management, it does not present any accurate method to predict the final degree of permanent disability in a workman.

The three-component theory of disability presented in chapter four has application in predicting disability if, in addition, we borrow from the world of physics. In depicting and measuring quantities such as force and velocity, physicists make use of vectors. A *vector* is a graphic representation of a measurable physical quantity having both magnitude and direction.

Vectors are depicted as arrows. The arrow's length is determined by the magnitude of the quantity shown. The direction in which the quantity is acting is indicated by the arrowhead. The following are examples of various vectors:

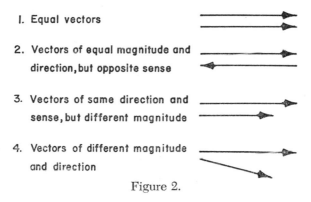

1. Equal vectors

2. Vectors of equal magnitude and direction, but opposite sense

3. Vectors of same direction and sense, but different magnitude

4. Vectors of different magnitude and direction

Figure 2.

The result of two or more vector quantities acting upon an object is the *resultant*. The resultant is also a measurable quantity and is likewise depicted by an arrow. To obtain the resultant of two or more vectors, they are summed.

The simplest manner of summing two or more vectors is by use of the polygon method shown as follows (Figure 3) in the case of three vectors.

The polygon method is merely the simple addition of one vector to another in a head-to-tail fashion. The resultant is obtained by drawing a line from the tail of the first vector to the head of the last vector.

Returning to the three-component theory, let's see how it is

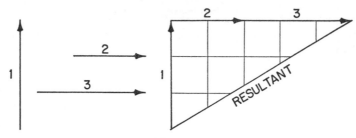

Figure 3.

possible to make use of vectors to predict permanent disability. By substituting the names of our three components of disability for the numbered vectors (Figure 4), we now have:

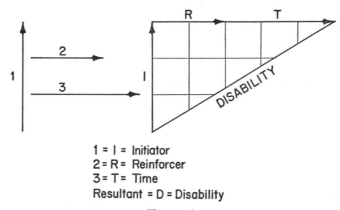

1 = I = Initiator
2 = R = Reinforcer
3 = T = Time
Resultant = D = Disability

Figure 4.

The resultant in this instance is disability. Since all three vectors and their resultants are measurable quantities, unit values can be assigned to each. Putting this to use, let's assume that any vector 1 cm. in length has a value of 1 unit. One unit in turn is equivalent to 5 per cent permanent disability of the whole man.

As an example, suppose we have an injured workman whom we determine has an initiator with an assigned value of 4 units, a reinforcer of 5 units, and a time period of 6 units (Figure 5). What is the predicted permanent disability?

See Figure 5 for the polygon method.

The predictable permanent disability in this example is 59 per cent of the whole man. This is an example reduced to its purest form, however. It is important to recognize that more

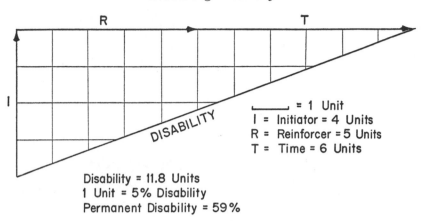

Disability = 11.8 Units
1 Unit = 5% Disability
Permanent Disability = 59%

Figure 5.

than one initiator, or reinforcer, may be an active force influencing the final permanent disability in any given workman.

With this basic example, let's now explore the effects each component has on disability. This is done by assigning variable hypothetical unit values to one component at a time.

In Figure 6 the initiator is shown as the greatest of the three forces present in this specific disability.

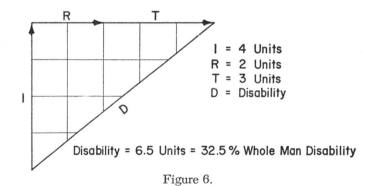

Disability = 6.5 Units = 32.5% Whole Man Disability

Figure 6.

Next, in Figure 7, the reinforcer is assigned a greater value than either the initiator or time factors.

Figure 8 shows the effect upon disability when time has a greater unit value than either the initiator or the reinforcer.

As to their individual effect upon permanent disability, either the time factor or the reinforcer has a greater relative value per unit than does an initiator. This is demonstrated in Figure 9.

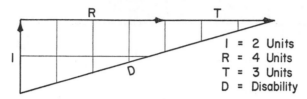

I = 2 Units
R = 4 Units
T = 3 Units
D = Disability

Disability = 7.4 Units = 37% Whole Man Disability

Figure 7.

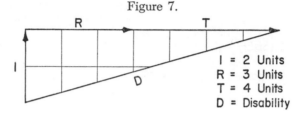

I = 2 Units
R = 3 Units
T = 4 Units
D = Disability

Disability = 7.4 Units = 37% Whole Man Disability

Figure 8.

Note that a 2 unit increase in the reinforcer (Figure 9B) increases the disability by 10 per cent when compared to Figure 9A. However, a 2 unit increase in the initiator (Figure 9C) increases the disability by only 3.5 per cent.

Let's turn now to a situation in which the initiator has the same unit value as Figure 9A. We shall also assume the time loss runs for the same period and has the same unit value. But, in this instance, those things reinforcing the workman's disability are weak and have a lesser unit value.

Now we are ready to shorten the time loss due to the injury (Figure 11). In this example, we leave the unit values for both the initiator and the reinforcer the same as they were in Figure 10.

In the final example (Figure 12), consider a situation in which the nature of the workman's injuries do disable him. However, during the entire recovery period, the perfect circumstances exist. No events take place that reinforce his initial disability.

Since the disability is not reinforced, the resultant vector representing disability is shown as a segmented line representing temporary disability. Time has a negative effect upon the initial disability in this instance because the workman's injury is healing. The initial disability also resolves because it is not reinforced during the recovery period. Finally, the workman

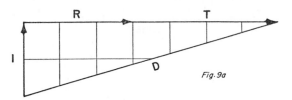

Disability = 7.4 Units = 37%

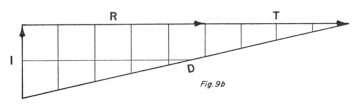

Disability = 9.4 Units = 47%

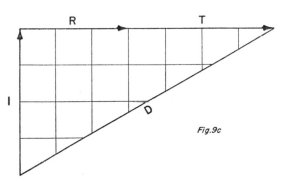

Disability = 8.1 Units = 40.5%

Figure 9.

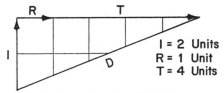

I = 2 Units
R = 1 Unit
T = 4 Units

Disability = 5.4 Units = 27% Whole Man Disability

Figure 10.

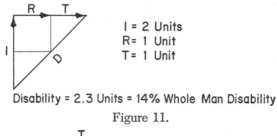

Disability = 2.3 Units = 14% Whole Man Disability

Figure 11.

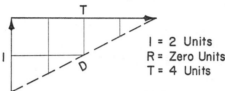

D = Temporary Disability = Zero Permanent Disability

Figure 12.

resumes working and living without any permanent disability.

These latter two examples should tell us something about disability and disabled workmen. It is possible to materially reduce or even prevent permanent disability arising out of work injuries.

Disability prevention is feasible. It is not purely a theory. A practical disability prevention plan is already in existence which is explained later in this text. Meanwhile, other matters need consideration.

RESEARCH

Detailed research is needed before the vector theory is applicable as a practical tool in claims management. Presently we do not have the knowledge required that allows us to implement the vector theory.

Identification of those factors that are significant in initiating and reinforcing permanent disability in the average permanently disabled workman is required. Once these factors are identified, it becomes necessary to assign a relative unit value to each.

To gain a little more insight into the nature of the research needed in relationship to disability and the vector theory, let's look at a present-day employer's reporting form for occupational injury. Currently, there are certain informational items on this form that are perhaps useful.

The need in using the useful information on this form is to identify which items are initiators, which are reinforcers, and what is the time frame we are likely looking at in a reported injury.

Age and sex are informational items that sometimes act as initiators of disability. It is possible for the age of the very young and the age of the older worker to serve as an initiating factor in the disability.

Even though the battle is on to erase discrimination, the mere fact that a workman is female can be an initiating factor. In addition, until a considerable change in attitude occurs, race is also an initiator of disability, but this information is now deleted from report forms.

The workman's address and occupation reported may even be considered as initiators of disability. It may be no other work exists in the locale for the specific workman reported as injured. The nature of his injury may preclude his returning immediately to his regular job. If there is no other job the employer can assign him to, he is disabled by this fact alone.

Whether or not a workman is hospitalized is indicated on the form. If he is hospitalized, this is an initiator of at least temporary total disability. And, of course, the nature of the injury reported may make it an initiator of disability.

Information on the report form may indicate that certain reinforcers of disability are present. For example, the attending physician may be known to the claim's manager. This physician may have a reputation of prolonging treatment of claimants under his care.

Additional reinforcers may lie in knowing who the employer is and the nature of the employer's business as reported. This particular employer may have a reputation for being slow in putting an injured employee back to work and this can reinforce a workman's disability.

The nature of the injury reported is an important informational item for the claims manager in a second manner. It gives the person managing the claim a clue as to the time component.

Anyone experienced in claims management has some knowledge of average recovery times for certain injuries. Therefore, if the specific injury reported is one that commonly requires several months for recovery, the likelihood of permanent disability is increased.

Using a theoretical example of a workman's claim, let's see how a claims manager might use this claims information to apply the vector theory to predict the degree of permanent disability likely to result from an injury to a workman.

Of course, at this point, we must assume that research has already established the relative values used in our example.

This is the claim of a twenty-eight year old male who suffers a fractured lumbar vertebral body. There is no cord or nerve injury.

This workman is employed by a large wood products company as a chokerman. He is also being treated by a physician well trained in the care of occupationally injured workmen.

As the person managing this workman's claim, you evaluate the information at hand and assign the relative values of the initiators, reinforcers, and time to predict the possible permanent disability that will result.

Because research has established these relative values, you find in this claim that the workman's age, sex and race have a zero value.

However, the workman's occupation has a unit value of 2 as an initiator. The fact that he lives in a very small community has a value of 1 unit. Since the employer is a large company, the possibility of re-employment is good and this makes it a very weak initiator of 0.5 units. An additional initiator is hospitalization with a value of 0.5 units. Thus, the initiators present in this claim total 4 units.

Turning to the reinforcers you find there are no obvious ones. The physician is competent and his patients tend to recover within the norms for their specific injuries. The employer is progressive and usually re-employs a workman in a job the workman can handle the moment the physician approves.

However, there is one more factor. This is the time factor. It appears to you that the workman will require three months to recover and this has a value of 1 unit. Therefore, the vectorgram will appear as seen in Figure 13.

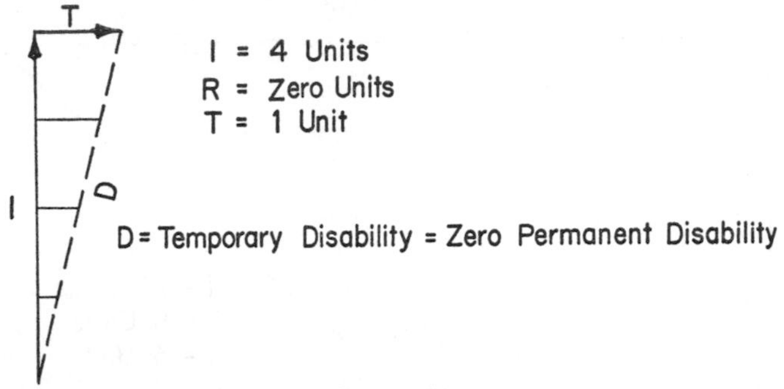

I = 4 Units
R = Zero Units
T = 1 Unit

D = Temporary Disability = Zero Permanent Disability

Figure 13.

In this claim you do not expect any permanent disability if the claim is managed properly because there is no indication that the temporary disability will be reinforced. This does not mean there will not be any physical impairment. There may be, but this is not disability.

Now, let's alter the same claim somewhat and then see what might be predicted. Let's make the workman a sixty year old timber faller working for a very small employer. In addition, the treating physician is known to be slow in releasing a workman for work.

In this instance, the age is an initiator having a value of 3 units. Occupation is 2 units. The small town is 1 unit. Size of employer is 1 unit and hospitalization has a 0.5 unit value. The total value for initiators is now 7.5 units.

In addition, there are some reinforcing factors. The physician's slowness to release a workman has a value of 2 units. The possibility of re-employment in the community is 4 units. The time for recovery is likely to be six to nine months because of the physician's tendency. Such a time factor has a value of 3 units. This vectorgram appears as seen in Figure 14.

Now you are confronted with a claim having a potential for a high permanent disability rating. However, it does not follow the workman has to have this large disability. It is dependent upon whether the claim is merely processed, or if it is managed.

The need is to identify those things which serve as initiators in most men, and the events which most often reinforce disability in injured workmen. Next we must determine the relative value of each. The research which permits this will also give

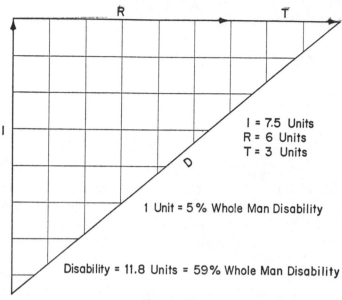

Figure 14.

us the ability to use this knowledge to manage each workman's claim.

To permit claims workers to disable a workman because they think only in terms of claims processing rather than energetic claims management, is wrong. It is, in fact, contrary to the philosophy underlying workmens' compensation law. However, we should not be misled by the belief that making more laws will correct this wrong. The correction comes through research that gives a greater insight into the meaning of disability, and a dedication of all men to the proposition that disability is preventable.

Chapter Six

DISABILITY PREVENTION

DEFINITION

D ISABILITY PREVENTION is the intentional effort to prevent or reduce disability manifested by the occupationally ill or injured workman.

A POSITIVE NOTE

Disability prevention is not synonymous with rehabilitation. It is, in fact, contrary to the meaning of rehabilitation. The meaning of rehabilitation is to restore. The meaning of disability prevention is to prevent. To prevent a disability in a specific workman leaves then no requirement to rehabilitate that workman.

As a function, disability prevention begins much earlier in the course of injury than do the present decisions and efforts to rehabilitate a workman. But we must keep in mind that physical and vocational rehabilitation play vital roles and are requisite within the overall structure of a proper disability prevention program.

In terms of need, the total numbers of ill or injured workmen requiring physical and vocational rehabilitation are small in comparison to the number of workmen needing assistance in the other aspects of disability prevention.

The thrust of disability prevention is pre-rehabilitation. Therefore, the effort begins and is concentrated out there, where the workman is, and where the workman lives, at the earliest possible moment after the injury occurs.

The workman is re-emphasized in disability prevention. In doing this, the workman is the focus of workmen's compensation law.

This does not imply that disability prevention prevails in an atmosphere of tender hearts and do-goodism. Nothing is provided a workman that is not already existent in workmen's compensation laws. Instead, disability prevention puts to use the great many benefits available to a workman which thereby prevent or reduce unwarranted disability.

Preventing the disablement of a workman is a positive approach to the occupationally ill or injured workman. In preventing any disability it is necessary to prevent those things leading to disability.

The injury is not preventable, only the disability. The injury or illness is already a fact. Properly functioning, the disability prevention effort centers upon the workman as a person. At this level disability prevention seeks to prevent:

1. Isolation of the workman;
2. Unwarranted fear responses in the workman;
3. Shifts in life styles and responsibilities;
4. Transference, or fault suppositions;
5. The adversary posture;
6. The compensatory mechanisms of time;
7. The loss of the work habit;
8. The summation of the emotional, economic, and social factors contributory to disability.

The prevention of these disabling factors comes about through the person-to-person contact. This aspect is mandatory if we hope to prevent, by any measurable degree, the disability in a workman. This face-to-face meeting first of all serves notice to the workman that he or she is recognized as a person.

The otherwise isolating effect of the illness or injury is interrupted by meeting the workman on his own ground. It is also mandatory at this first meeting with a workman that his rights and benefits under workmen's compensation law are made very clear.

Knowledge of these rights and benefits alleviates many unwarranted fears presently founded in workmen by gossip and piecemeal facts supplied through outside sources.

A clear understanding of workmen's compensation law by a workman, as that law pertains to his rights and benefits, is a positive step toward preventing an adversary attitude later.

Personal contact is also a means of keeping the workman's responsibilities and the likelihood of resuming work before him, and the probability of a shorter recovery period which thwarts the workman's acquiring the disability habit by adjusting to his plight.

As an example of the beneficial effects resulting from a positive disability prevention program, let's examine the claim file of a workman in the state of Oregon. This first example is not one of disability prevention in its true sense. But, as the story is examined in this claim file, we will see how an early positive disability prevention effort should have taken place.

A disability prevention service coordinator became involved with this workman very late in the claim. In spite of this he brought this workman's problems to a rapid and successful conclusion. Had the coordinator not been involved, there is every indication that this workman would have remained severely disabled.

This particular workman, after twenty-three years of employment with a large utility company, injured his low back on June 16, 1969.

At the time of his injury, the workman worked as a power equipment operator. His job also required some heavy manual labor and pole climbing in addition to operating the power equipment.

Upon injuring his back he went to a chiropractor where he received eight treatments. He was then reportedly improved and released for work. But, the chiropractor indicated in a report that this workman had a herniated disc, even though he responded to treatment.

After the workman resumed working, his back condition gradually worsened. He then saw an M.D. who prescribed medication, and he continued to work until three weeks prior to admission to a hospital on October 3, 1969. This was now nearly four months after his original injury.

During hospitalization a neurosurgeon saw this man and subsequently performed a laminectomy. At surgery, the neurosurgeon removed a large intervertebral disc from between the fourth and fifth lumbar vertebra.

Postoperative recovery was reportedly good, according to the claim file, that is. However, something went awry. The workman didn't resume working as we might expect from the information contained in the claim file up to this point.

Was the workman now *unmotivated* to return to work? Perhaps. But we must remember that this man did work for more than three months after injury although his back became progressively worse.

Whatever the reasons are, no clue exists in the claim file indicating the causes of this workman's prolonged disability. But there is a clue of beginning conflict. In mid-November, while recuperating from his surgery, the workman wrote a letter asking the insurer about his compensation check.

The workman had not received a compensation check for his time loss during his entire hospital stay, and he indicated this was causing him problems in meeting his bills. With this the insurer sent the workman the compensation due, but without any explanation to the man.

Eleven months from the date of injury things were still bumping along in the claim file and the workman was still on time loss. A post card from the workman states, "I would sure like to get back my old job."

The insurer's reaction to the workman's post card was zilch, as though the man didn't exist. So, two weeks later the workman visited the insurer's office. He inquired as to when the insurer would close his claim and he was informed by the claims adjuster that they could not close a claim without the treating physician's recommendation.

Three weeks after this visit, the insurer wrote to the treating physician asking if the workman's condition was ready for claim closure.

The treating physician meanwhile had examined the man, but failed to respond to the insurer's request for a report. Whereupon, the insurer notified the workman they were stopping his compensation payments unless they received the physician's report.

The workman replied, reporting that he went to the physician's office where he pleaded with the office nurse to send the insurer the report.

The insurer finally closed the claim on October 5, 1970, and the workman received a permanent disability award of $5,225, to be paid out on a monthly basis. But, without explanation, the insurer withheld $85 from the workman's first check because of prior overpayment. This upset the workman.

The workman immediately wrote a letter of inquiry asking why the insurer had shorted him. He indicated he was already in financial trouble. The insurer replied, explaining a prior overpayment in temporary total disability benefits.

One year and one month after claim closure, the insurer referred the workman to the Division of Vocational Rehabilitation. This was now two years and five months after his injury. The workman was still totally disabled, although the treating physician had originally believed the workman could resume working when he recommended claim closure more than a year before.

The workman informed the vocational rehabilitation counselor that he preferred returning to work for his old employer. Thereupon, the counselor advised the workman to contact the personnel manager of the company. There is nothing in the claim indicating that the counselor made any effort to assist in this by contacting the employer.

Vocational counseling continued for the next seven months. A reason given for this by the counselor was the workman's poor attitude. However, the counselor did indicate that he believed the workman could work as a meter reader, and he advised the workman to see the employer about such a job.

The workman found two possible jobs as a meter reader with the company. One required that the workman move from the community in which he had spent nearly his entire life. This meant selling his home of thirty years which he now owned. He didn't take the job because of this, and because of his family's desire to remain in its present community.

Another nearby district personnel manager then offered the workman a job as a meter reader. However, this personnel manager demanded that the workman first sign an agreement to read 450 meters daily.

The workman refused this job for fear he couldn't meet the demand. In addition to being unsure of his back problem, this workman is also a wearer of trifocal glasses. Between the two problems, he believed he would be unable to read 450 meters daily.

During five months of assistance and counseling by the vocational counselor, the workman lost faith in their guidance. The workman somehow learned of the service coordinator and the new disability prevention program implemented by Oregon's Workmen's Compensation Board.

Therefore, one July day the workman walked into the service coordinator's office asking for help. The service coordinator heard the workman's story and set to work on the problem. After a twenty-minute telephone conversation with the chief personnel manager for the utility company, the personnel manager agreed to put the workman back to work.

Some three years after his injury, this workman began working for the company as a meter reader on the last day of July 1972, in the same district where he worked prior to his injury. Also, the company hired him back at the same wage he had made in his job as equipment operator.

Perhaps the story of this workman and his prolonged disability appears to make the service coordinator into a hero. In less than half a day of effort, the service coordinator had this man a job with the same employer. However, the service coordinator is not a hero. Instead, he took advantage of the circumstances.

First of all, the workman came to the service coordinator asking for help in finding a job. This is a plus factor for the coordinator to use. Secondly, the workman's compensation award benefit payments were at an end. The workman was now faced with the possibility of seeking help from welfare, which he couldn't bring himself to face.

The service coordinator likely used this latter factor as leverage with the employer. He probably explained to the employer that the workman might now file a claim for aggravation of his injury if he didn't find a job, rather than accepting welfare assistance.

In turn, the employer probably considered the fact that the workman might seek a permanent total disability through his claim of aggravation. The possibility existed that those proceedings might end in favor of a workman who has an unemployment record lasting three years after an injury.

By good fortune this workman preferred working to loitering and accepted the employer's job offer. Happily, the workman is still employed as a meter reader after more than a year. Unfortunately, he could have resumed work much sooner had someone taken an interest.

True, the insurer did refer the workman for vocational rehabilitation. But it was a long, long time after the injury. Probably by then the referral was made only because the carrier finally realized the potential possibility of a permanent total disability claimant in the offing.

Perhaps the vocational rehabilitation counselor's counseling was valuable. His actions were weak. Had the counselor taken the same actions as the service coordinator, the workman likely would have resumed working five months sooner.

While the results in this claim are fortunate, the delay is tragic. The personal losses to this workman, and the general losses to society, due to prolonged disability as seen in this situation are unnecessary and preventable.

Early action in preventing disability doesn't come easy, even today. Insurers, their employees, and others are not yet trained to think in terms of disability prevention. Being human, they resist changes in claims management and thinking.

The next example shows how this resistance to thinking in terms of disability prevention contributes to disability. In this example, the insurer had full knowledge of the disability prevention program in existence. But the insurance company employee managing this claim is seen resisting any change in methods or thoughts.

This is the claim of a forty-nine year old man who injured his low back. His treating physician referred him to a neurosurgeon soon after injury. The neurosurgeon did a thorough examination including a myelogram. Another neurosurgeon also saw the man in consultation.

After conservative treatment for four months, the neurosurgeon concluded the man's problems were due to emotional disturbances. Whereupon the neurosurgeon recommended claim closure on the basis that no other treatment would benefit the man.

The claim was closed, awarding the claimant $2,240 as a permanent partial disability award. However, the claimant continued to have constant pain and made no attempt to return to work.

Because of persistent pain, the claimant next came to the insurer's office requesting permission to see a chiropractor. The interviewer for the insurer informed the claimant his claim was closed and they could not pay for any further medical treatment.

Two months later the claimant sought medical treatment by a chiropractor. The chiropractor mailed the insurer a narrative report of his findings and recommended further treatment.

The insurer replied to the chiropractor asking that the claimant be referred back to the neurosurgeon who first recommended claim closure. Meanwhile, the insurer assigned an investigator to the case.

The investigator's report revealed that the claimant would rather be working "than hanging around the house all day." In fact, the claimant reportedly submitted several job applications to various employers. There was no response from those employers and the claimant believed them to be "gun-shy" of a man having a bad back.

In the investigator's report, the claimant also reported definite improvement in his back since going to the chiropractor. His back and leg pain was no longer a steady pain. Although he did experience some recurrences of pain, it was not as often since he began the chiropractic treatments.

The workman also told the investigator he was now able to do more physical work about his home. In addition, he informed the investigator that he could see no point in returning to the neurosurgeon who had been unable to bring him relief from his pain in the beginning.

In spite of this information in the investigator's report, the claims adjuster then decided to refer the claimant to an orthopedist for an examination and opinion.

The orthopedist noted in his report the improvement the claimant obtained from chiropractic treatment. But the orthopedist could see no point in reopening the claim for treatment. The orthopedist considered the claimant fit to work. However, he made no recommendation as to the type of work.

On the basis of the orthopedist's report, the claims adjuster denied the chiropractor's bills and refused to reopen the claim. Meanwhile, other events were occurring.

During the time of this struggle between the claimant and the claims adjuster, a positive action took place. The chiropractor, knowing of the disability prevention program operated by the Oregon Workmen's Compensation Board, referred the claimant to a service coordinator. The service coordinator took immediate action and one week prior to the time of the insurance claims adjuster's last action, the claimant began working.

The service coordinator in talking with the claimant found that this man had other talents. The coordinator was able to place the claimant in a job as an assistant manager of a fraternal club. The claimant began his new job at a salary of $600 per month. This was an increase of $200 a month over his earnings prior to his injury when he worked as a laborer.

After returning to work and after receiving the denial letter from the claims adjuster, the claimant visited the insurer's office. This time a claims supervisor talked with the claimant. We now find in the claim file instructions from the claims supervisor to the claims adjuster to pay the chiropractor for the services rendered.

The supervisor, in his instructions to the claims adjuster, advised him that the claimant was ready to hire an attorney. The supervisor decided it was wiser if they agreed to pay the claimant's medical care.

The claimant is still working and no further treatment is reported in the claim six months later.

The period of total disability for this claimant lasted ten months. The disability time could have been shortened had the claims adjuster for the insurer acted promptly in the beginning, or at any time during the ten months. Instead, the actions of the claims adjuster very nearly precipitated a law suit.

Sometimes the act of disability prevention is surprisingly easy. This is the situation in this next example of the deputy sheriff. This claimant injured his back, and the treating physician referred him to a neurosurgeon. The neurosurgeon performed a laminectomy, removing a herniated disc just eight days after the date of injury.

The insurer requested the assistance of the service coordinator one month after the claimant's surgery. At the initial contact, the service coordinator found the claimant eager to return to work and seemingly recovering very well from the surgery.

However, the claimant's employer was reluctant to let the claimant return to work until the neurosurgeon advised them the claimant could work. The employer made no inquiry of the neurosurgeon. Seemingly, the employer was content merely to wait for such advice to originate with the neurosurgeon.

When the employer was contacted by the service coordinator, the employer agreed to modify the claimant's job and allow him to resume work. The neurosurgeon also agreed to this when the service coordinator sought his advice.

As a result of the service coordinator's relatively easy efforts and the agreements reached, the claimant resumed working in a modified job six weeks from the day of the laminectomy and disc removal.

This claimant who suffered a back injury requiring major surgical correction lost a total of seven weeks from work. This time loss would have been longer had the insurer delayed the referral to the service coordinator, and had the employer not been asked to cooperate by modifying the claimant's job.

All parties to this claim were cooperative. The service coordinator served as the cohesive agent in this claim by bringing the cooperative spirits of the insurer, the physician, and the employer together.

Because disability prevention is an act of preventing, the effectiveness of this concept is dependent upon changes in present-day methods and practices in both claims management, and the total care of injured workmen. To change these methods and practices within the respective compensation systems requires varying degrees of effort.

The means and mechanisms to prevent disability exist today. It does require that clear-minded men, with foresight, work together utilizing the means at hand.

LEADERSHIP

In many states disability prevention is already in practice, but in a fragmentary manner. For example, the Liberty Mutual Insurance Company has employed nurses to act as service coordinators for several years. In addition, the company maintains its own rehabilitation center in Boston where certain injured workmen are treated.

Many large companies and corporations maintain medical services departments and practice many of the concepts of disability prevention. Their efforts bring definite benefits to their employees by minimizing time loss and unnecessary, degrading disability. Wisely, these organizations view such practices as protecting their investment in the specific employee.

For a statewide disability prevention plan to be effective, it requires a single leadership. The bulk of employers in any state cannot afford the services provided the employees of larger companies, nor do most insurers supply such services.

THE ADMINISTRATIVE AGENCY

The administrative agency responsible for the functions of a state's workmen's compensation law is best suited for leadership in disability prevention. Ideally, the state agency administering the law is an independent agency which exercises the regulatory powers but does not in fact enter into the insuring of employers such as we find with the state funds.

If a state cannot separate the administrative and regulatory powers from its insuring agency, it remains essential that the disability prevention plan functions in an atmosphere of neutrality. In all cases it is better to establish the disability prevention division as an autonomous unit within the administrative agency.

Nonpartisanship is mandatory for an effective disability prevention plan in any state. This neutrality of the disability prevention division cannot be overstressed. One of the values in this neutrality lies in reducing the risk of creating adversary attitudes in the workman toward the workmen's compensation system.

The disability prevention division functions in many ways. The details of these functions are described later in the text. However, there is one function that is appropriate to discuss now.

The work of the disability prevention division must never occur as an offshoot of a vocational or physical rehabilitation division. Conversely, vocational and physical rehabilitation are in themselves responsibilities of the disability prevention division.

The acts of disability prevention occur as pre-rehabilitation actions at the local level where the workman resides. When physical and vocational rehabilitation are needed by the workman, the quality and the promptness with which these services are provided is critical to the outcome. Therefore, one of the functions of a disability prevention division is to coordinate the timely provision of the highest quality rehabilitation available.

Presently, in most states, the decisions and the recommendations for either form of rehabilitation are usually left to the treating physicians and the insurers.

This method of decision making is not satisfactory. It leaves too much of the responsibility to parties whose interests may even be self-serving. As a result, the decision is either not made or is unnecessarily delayed.

A study made of the permanent total disability claims arising out of low back injury claims for the preceding five years was completed in 1969 by Martin.[15]

In Martin's study, 144 workmen were permanently totally disabled in the preceding five years because of a low back injury. Of the 144 workmen, 56 per cent were never referred to a physical rehabilitation center for care or treatment.

Equally disturbing is the delay seen in which the remaining workmen were referred to a physical rehabilitation center. This group represented 44 per cent of the workmen in the study. In this group, the average workman was not referred for physical rehabilitation until eighteen months after the date of his injury.

In this same study the findings of referral for vocational rehabilitation were even poorer. Referral for consideration of vocational training was limited to 45 per cent of the claimants. The time lag for referral from the date of injury was one year or more in 80 per cent of those claimants referred.

Although late referral is better than no referral at all by insurers or treating physicians, late referral for rehabilitation unnecessarily disables the workman in either a temporary or permanent manner.

Experts in the field of rehabilitation constantly decry this practice of late referral for rehabilitation. Although we appear willing to acknowledge their expertise, we fail to hear their pleading. Prompt referral for vocational and physical rehabilitation when indicated is paramount to disability prevention in that specific workman.

Thus, one of the responsibilities of the disability prevention division is the coordination of rehabilitation to assure its prompt application. Within this responsibility is also an obligation to assure that the form of rehabilitation provided meets the highest professional standards.

To provide physical rehabilitation services that are inadequate and of poor quality is as negligent an act as permitting an untrained surgeon to perform open heart surgery in his office.

It is as important to perform the procedures of physical rehabilitation in a proper and well-equipped setting, utilizing well-trained personnel, as it is to perform surgery under highly adequate circumstances. Much is at stake in both instances.

A good physical rehabilitation facility goes beyond the common concepts of physical therapy, and the usual occupational therapy commonly provided in many physical rehabilitation centers.

An effective physical rehabilitation facility has both physiotherapy and occupational therapy, of course. But, it does little good to restore use to an injured part only to find that the workman is in such a lousy general physical shape he can't hack a day's work.

Conditioning the workman's whole body and maintaining the whole workman in a state of good physical health is mandatory in any physical rehabilitation effort. An adequate facility provides for this type of physical conditioning.

Finally, a good physical rehabilitation facility also toughens the workman to the expected work situation. It can be called "industrial therapy" or any number of names. Basically, it is the act of putting the man back to work while he is at the rehabilitation facility.

While physical and occupational therapy are important treatment forms, industrial therapy also has a function extending beyond merely toughening the man. Physical and occupational therapy are mostly forms of *doing for* the workman. Industrial therapy causes the workman to *do for himself*.

The industrial therapy department of a physical rehabilitation facility has a variety of both indoor and outdoor work stations in which the workman is conditioned to work. While it is impossible to simulate all classes of jobs, it is possible to condition and evaluate his ability to work a full or part day in what is likely to be his major activity in his regular work when he is fully recovered.

Industrial therapy, however, is not to be confused with industrial training. Industrial therapy in a physical rehabilitation setting is a treatment form. Industrial training rightfully belongs in the realm of vocational rehabilitation.

However, industrial therapy can also be used to evaluate a workman's abilities in the area of proposed industrial training. Thus, industrial therapy is additionally useful for vocational evaluation purposes.

When a workman completes his treatment in a full-service physical rehabilitation facility, the work abilities and the work stamina of the workman are known to all, including the workman. This in itself is good psychotherapy, or applied psychology at its best.

Vocational rehabilitation of injured workmen is ideally controlled within the disability prevention program of the workmen's compensation administration agency.

This does not mean that vocational rehabilitation of claimant workmen cannot be under the supervision of the designated vocational rehabilitation division (V.R.D.) of the respective states. This separation of function found in the states creates a critical area in any disability prevention program.

The problems in this division of function arises because the vocational rehabilitation division has other obligations. It must concern itself with the vocational rehabilitation of all segments of society. The claimant workman may not have a priority.

In disability prevention the necessary retraining of any injured workman must have top priority. If the injured workman is forced to wait for long periods of time while a vocational counselor plans a training program, the disability prevention attempt is diluted or even destroyed.

Therefore, if the task of managing vocational training is assigned to V.R.D., it is better if a contract exists between the two agencies which assures the special assignment of V.R.D. counselors whose only task is to work with workmen's compensation claimants.

However, one responsibility of disability prevention is to minimize the importance of the more formal types of vocational retraining. It is likely that this thought is sacrilegious to many stalwarts who view the upgrading of knowledge and skills of vocational rehabilitation clients as sacrosanct.

There is another way though of looking at this matter of vocational rehabilitation. Upgrading the person may be of value in some instances. On the other hand, simple, straightforward, on-the-job training is a preferable solution for most workmen, and there is reason.

The reason relates to the educational history of the average workman. Most workmen limit their education by their own will. They are in a sense more practical people than those persons who seek considerably more education.

The workman who limits his education does so because it is the *here and now* that has the most meaning. To such a person the future is a nebulous, unclear gap in time. It is difficult for them to relate the *now* to the future.

Because it is hard for this person to relate to the future, formal education makes little sense. That spelling, and grammar, and mathematics, or physics, history, chemistry, and the other subjects taught may lead them to a better living in the future is unreal to this person.

It is less than sensible to place a person of this nature in a similar schooling situation under the auspices of vocational rehabilitation. Even though a specific workman may have the mentality to complete the educational plan designed, it is still contrary to his nature.

But work has meaning. Work is the here and now for this workman. Building, assembling, shaping, accomplishing, and earning, occurs during work. Such work is visible, touchable, and sensible to this workman.

Therefore, on-the-job training is a better solution for most workmen because the average workman has already served notice on us that gainful work has more meaning than the theories of education and the ethereal future.

In addition, it is not the task of workmen's compensation to upgrade the lives of all claimants through vocational rehabilitation. Except where no alternative exists, educational upgrading is the responsibility of the whole of society and is another matter.

The purpose of vocational rehabilitation within workmen's compensation is to make the workman fit for a gainful occupation. It is perfectly proper to accomplish this mission through the most reasonable means available involving the least possible effort.

This is the sort of leadership a workmen's compensation agency and its disability prevention division must develop. The agency must establish in the minds of other parties the expertise the agency has in matters of workmen's compensation.

In addition, leadership requires an agency to define the roles of the other parties to a claim. Certain guidelines are required to keep the parties on course. This, though, is a matter of thought and education, not law making.

In assuming this leadership, the agency and the disability prevention division must keep in mind the importance of encouragement to the other parties. It is not a function of the agency to discourage insurers and employers in their own practices of disability prevention. Rather, it is part of the agency's role to encourage and coordinate such practices.

SUMMARY

Disability in a man or woman is as preventable as many diseases. Disability prevention is within the grasp of all states. Many benefits accrue to a state dedicated to disability prevention. One such benefit is the state's ability to maintain a more viable work force. To arrive at this requires preventive care by the care team in workmen's compensation.

Chapter Seven

THE CARE TEAM

Workmen's compensation law provides for a primary care team which rarely functions properly as a team. This team is composed of the treating physician, the insurer, and the employer.

It lies within the ability of this team to prevent much of the disability arising out of our present-day occupational illnesses and injuries. It requires the dedication of each team member to the proposition _that no man benefits from a disability._

With this as a beginning, each care team member is obligated to evaluate his own role. In doing this, each must also recognize the expertise of his fellow members. In addition, all parties to the care team should accept the criticisms of faults believed to exist in their respective delivery systems.

The discussion in this chapter is a consideration of the faults seen in the delivery systems of the medical profession, the insurers, and employers. Since they are primarily the major parties to a workman's claim, recognition of these faults is pertinent to evaluation of their roles in disability prevention.

TREATING PHYSICIANS

The physician's traditional role in treating occupationally ill and injured workmen is due for alteration. The treating physician's role is being challenged in workmen's compensation today. This is a reflection of the complex nature of interrelationships developing in today's society.

Increasing emphasis on the obligations of the insurers, the employers, and the agencies in the care of the ill or injured workman is obvious. Changes in social thinking and court decisions concerning disability are also strong factors in this shift of roles. Care of the workman is no longer an isolated prerogative of the medical world.

Care is no longer a term with the singular connotation of medical treatment in workmen's compensation. Care extends beyond preserving the workman's physical life. Care now involves the preservation of a workman's economic life, his work life, and the continuation of his dues-paying membership in society.

Therefore, the treating physician is finding himself joined by these other entities in what is essentially a *total care team.*

Because many treating physicians do not fully comprehend these changes, the doctor is not always a willing member of the care team. Physicians are naturally reluctant to give up their traditional role. They tend to view this as third party intrusion in the care of the patient. However, the day of the old treating doctor being all things to all men is over. The demands upon today's physicians detract from the time spent in otherwise caring for the patient.

Take the simple matter of home calls. They are nearly history in themselves. The physician of yesteryear learned much by making home calls. Today, this sort of knowledge about his patient is becoming a rare privilege for a busy physician.

More and more treating physicians know less and less about the person they treat. The decrease in family practices and the increase in specialization is one of the limiting factors in the ability of a doctor to know his patient.

For instance, how many orthopedists, while treating Joe's back, ever meet Joe's wife? Has the orthopedist ever helped her through a nervous tizzy, or discovered what a shrew she is, or learned perhaps the fact that she's a doomsdayer?

How many orthopedists, knowing not one thing about Joe's home and family life, find themselves led into the surgical suite by Joe because he isn't responding to their conservative treatment?

How are these orthopedists going to know? Their time is limited. Their offices are crowded. In addition, the orthopedist is not likely to seek out such information by making a home call. Nor is he likely to have an opportunity to treat the patient's wife or family in a manner that gives him a deeper insight into the patient's life.

In addition, it is no longer acceptable practice for the treating

physician to concern himself only with the healing of the workman's wounds. It is not sufficient practice to heal the workman
and then let him find his own way back into a gainful life.

Changing social concepts and social responsibilities demand
that treating physicians join in as care team members. One such
reason why physicians must cooperate and participate as care
team members is this matter of prolonged disability.

The length of time a workman is disabled by injury is increasing. This is occurring regardless of our glowing accounts of
sophisticated medical care. From this, it is evident that modern
medical care alone is not the whole determinant in whether or
not a workman recovers. The search is on to determine and to
deal with all the factors contributing to a workman's disability.
Medical care, as it is applied to the care of occupationally ill or
injured workmen, is now under scrutiny of the providers of care
in workmen's compensation law.

The parties to the provision of medical care for claimant workmen have a lawful obligation to evaluate the problems presently
existing in the medical care delivery system as it functions in
relationship to workmen's compensation.

Presently there are three basic areas that the medical profession itself needs to bring under close scrutiny by asking itself:

Who Cares?

Traditionally, the treating physician cares for the workman
patient. We might ask ourselves now if this tradition should
continue.

This questioning of the tradition is the result of the changes
in the practice of medicine in the past twenty years. The age of
specialization is now, and the treating physician's role as practiced today is limited.

Modern medical care rightfully boasts of astounding advances
that otherwise might leave the workman hopelessly crippled by
the occupational disease or injury. However, these workmen are
not commonly the problem people.

The majority of workmen sorely disabled in workmen's compensation do not suffer horrendous, life-threatening occupational
illnesses and injuries. Furthermore, there is no evidence that

modern medical care is effectively reducing the number of severely disabled workmen or the degree of their disability.

In fact, there is reason to believe that modern medical care contributes in some degree to the disability in a large number of workmen. Psychiatrists and psychologists recognize this, but the rest of the medical profession seemingly chooses to ignore their warnings. There is ample reason to believe both the psychiatrists and the psychologists, if we give some thought to workmen and their disabilities.

The average workman we see having a large degree of disability today is disabled in part by the atmosphere prevalent in modern medical care. Because this workman is the kind of person he or she is, he does not respond well to care provided in an emotionally sterile environment, or to the impersonal relationships we see so much of in today's medical environment.

These workmen do not respond as well to the tinkerings of a parts doctor. Such a person needs whole treatment by a whole doctor. It is wrong for the medical profession to foist piecemeal, highly specialized professional care onto any workman needing less modern technology, but more care.

The need and also the responsibility for the specialist is to recognize his own deficiencies in the total care, or treatment, of a specific workman manifesting disability.

The correction of this problem comes about if first the specialist admits to his own limited role in the care of a workman. Secondly, it requires the specialist to assume a role subservient to a physician experienced in managing the care of the whole man.

This is a break in traditional roles played by both the specialist and the general physician. Heretofore the general physician's role has been commonly subordinate to that of the specialist.

To reduce the disabling effects of modern medical care requires the general physician to assume the role of leadership. In this role, the general physician then has the basic responsibility for the proper utilization of consultants. It is the general physician's chore to determine whether or not a consultant's advice is in the best interest of a specific workman.

Additionally, this role also obligates the general physician to function as a care team member with insurers, employers, and agencies. In turn, the latter three must make contributions that are vital in assisting this physician's medical decisions in the best interest of the workman.

The general physician is better suited for the care of claimant workmen than the specialist, if the general physician accepts his responsibilities of decision making and care team membership.

None of this is meant to demean those physicians who are not general care physicians. It may bend some egos. However, it is the workman that is being treated, not the physician. Therefore, it is time that the medical professions recognize the problem and determine who it is that cares for the workman.

Utilization

Over-utilization of medical care clearly exists in workmen's compensation claims. To what degree is unknown, but it may be more rampant than is generally believed.

Steps to assess the degree and the causes of over-utilization are an obligation of the medical profession. It is time the profession assumes its obligation of leadership in this matter and enlists the aid of insurers, employers, and state agencies in evaluating the problem.

Because over-utilization of medical care is not presently openly discussed in workmen's compensation, it does not mean that it is not being considered. Utilization is closely guarded in those patients entitled to medicare, and the private physician can soon expect this same turn within workmen's compensation insurance.

With medical care costs mounting, more rigid control of utilization in medical care is the one area whereby some cost control is possible. Since the National Commission is calling for closer supervision of medical care by workmen's compensation agencies, utilization review is at hand.

Although cost control may be a factor causing the other parties to investigate the matters of over-utilization, there is an additional harmful aspect to over-utilization. Prolonged treatment, like improper treatment, contributes to a workman's disability.

In spite of the medical profession's claim of its self-policing policies through peer review, the consumer's anxiety in workmen's compensation is growing.

This growth in consumer unrest stems in part from the insistence by the medical profession for payment of a usual fee rather than the use of fee schedules. Previously, fee schedules controlled somewhat the total cost by limiting both the dollars paid and, to some degree, the amount of treatment.

With the increasing acceptance of the usual fee concept, the consumer is now looking harder at utilization. Over-utilization will be controlled in workmen's compensation by either the medical profession or the consumer. At this time the medical profession still has a choice.

Regardless of choice, it is the individual physician's professional obligation to assess his treatment methods, to eliminate over-utilization in all aspects of the care he provides his patient. The physician who prevents over-utilization prevents unwarranted disability in his patient.

Work and Expertise

The average treating physician is not expert in the world of work. Excepting the few physicians employed by industrial firms, the average physician is ignorant of the goings on in the big world outside his office.

The evolution of the physician in this role as a supposed expert in job evaluation is understandable. Because physicians have expertise in the care of injuries and illnesses, they are naturally expected to be the ones who determine when the workman can resume work.

However, in spite of the efforts of industry to acquaint physicians with various jobs, the average physician for a variety of reasons never responds to the opportunity. It is unfortunate, but true. Thus, a large hiatus exists in the average physician's expertise in the work world.

Because of the technological changes occurring in industry which require less muscular effort and less motion by a workman to perform the task, too many treating physicians issue uninformed statements restricting the workman's return to work.

Unless the physician has a special interest in keeping pace with the changing nature of occupations, he should not pose as an expert.

The work requirements of jobs today are very different from those twenty years ago. For example, in the West, working on the green chain in a sawmill is traditionally considered a rigorous, demanding job requiring a strong back. This is not true in some automated sawmills, however.

In certain mills a man working on the green chain is now an instrument operator. This man sits on a comfortable stool, and with the aid of a television camera sorts and stacks the lumber by merely pressing a few buttons on a small black box. This is the labor of a modern green chain. But, how many western physicians know this?

The obligation of the uninformed treating physician, as a care team member, is to admit his deficiency. An alternative is to request specific work information. He can also advise others of the workman's physical abilities. Together the team can match the workman to a job within his physical limits.

A physician who is not fully informed and arbitrarily claims that a workman is incapable of an available job contributes a great deal to the disability of the workman. In addition, such practices by a physician may border on malpractice.

Basic Responsibilities

With these criticisms in mind, let's turn to three specific obligations a treating physician has as a care team member in providing medical care to the claimant workman.

1. *Acute care and convalescence.*

   Treatment of the occupationally ill or injured workman through the acute phase and convalescence is the natural role of the treating physician. This treatment should continue until the workman is medically stationary, but not beyond.

   Consultation should be used on a judicious and timely basis. The workman should be viewed and treated as a whole man, not on a piecemeal, specialized parts basis. Thus, the treating physician, whether a specialist or not, is responsible for the whole man.

While convalescence does not require the attentiveness of the acute phase, it does require active care. Convalescence is not a period wherein the patient is relegated to insulation or isolation from life.

Convalescence is a period of regrowth and re-absorption into life. Although a workman may not necessarily require frequent direct attention from the physician, he does require direction from the physician.

While the treating physician can supply all the needs of medical convalescence, the physician is also obligated to inform the care team as to other possible needs of the workman.

This does not mean the treating physician is expected to serve as an investigator in determining those needs. It means he is capable of foreseeing needs such as psychological counseling, or physical and vocational rehabilitation.

2. *Rehabilitation*

There is no excuse for a treating physician who delays initiating rehabilitation of a workman when it is required in the recovery process.

Even less acceptable is the physician who undertakes the physical rehabilitation of a workman simply because that physician purchased some therapy equipment. The application of heat through the use of diathermy, ultrasound, or sundry other devices by untrained employees posing as therapists is not physical therapy.

Nor are instructions to the patient to putter about the house making use of the injured part suitable occupational therapy, where such therapy is needed. Neither of these inept attempts qualifies as physical rehabilitation.

Evaluating the need for both physical and vocational rehabilitation is always a required part of the acute care provided to a workman by a competent physician.

It is also the responsibility of the physician to initiate through the care team the required rehabilitation at the earliest possible time during the recovery process.

Extensive and prolonged physical rehabilitation requires that it be undertaken in a facility dedicated to that purpose, a rehabilitation center. If the situation requires that a treating physician relinquish his care of the workman to another physician to provide timely, high quality physical rehabilitation, it is the treating physician's obligation to willingly make the referral.

While vocational rehabilitation is not within the physician's domain to direct, it is a partial responsibility of the physician to his patient. As a member of the team, the physician has the first opportunity and an obligation to assess and recommend needed vocational rehabilitation.

3. *Medical Reporting*

In reviewing claims of problem cases, the most outstanding feature is the generally poor quality of medical reports issued from the treating physicians' offices.

This lack of proper information is undoubtedly a contributing factor in the creation of such claims. While this whole business of communications cuts both ways, one of the disciplines of the medical profession is the production of adequate medical records.

There are found in claim files medical reports which, if read, will make anyone question the doctor's good sense. Such poor reports are numerous in the claim files.

The California State Compensation Insurance Fund [15] is increasing its efforts to coordinate the care of the claimant with the treating doctor. This change stems from studies conducted by the fund.

The studies in California reveal that there is no predictive value in the initial diagnosis reported by the treating physician in selecting those workmen likely to evidence prolonged disability, or marked degrees of permanent disability. Nor do the initial predictions of the workmen's return to work, as given by the treating physician, have any validity.

Although people knowledgeable in workmen's compensation affairs have long expressed these same thoughts on the reliability of the initial diagnosis and predicted time loss by treating physicians, the California study confirms these facts.

Surely professional people as disciplined as Doctors of Medicine or Osteopathy are capable of a more reliable initial diagnosis and prediction of time loss than is evidenced by their sorry records in workmen's compensation.

A physician renders a valuable service to the workman when an adequate medical report is issued by that physician. Failure to do so also represents a disservice.

Medical reports should always be in professional language with no attempt by the physician to speak in lay terms. The reports should contain the same pertinent information as is normally contained in the proper interchange of information during consultation.

Resting upon the medical reports are important decisions for the workman such as continued medical care, continued time loss payments, rehabilitation, and the very important matter of disability.

Not all the fault, however, lies at the feet of the physicians. Many medical reports, perhaps the majority, are made through pre-printed report forms.

These forms neither ask for information nor provide space for the physician to give information that depicts the patient in the proper light to always avoid a problem. While routine forms suffice in the reporting of trivial injuries and illnesses, they are not sufficiently informative in the more complicated claims.

Arguments exist on both sides in this matter of medical reporting. Physicians complain they are smothered in requests for reports. Insurers claim it is too difficult to obtain a report from physicians. Both sides are right.

However, it is past time for the medical profession and

business to embark upon the solution to effective medical reporting. Meanwhile, the conscientious physician will strive to produce useful reports for the benefit of his patient.

INSURERS

The role most critical to the final outcome in any specific claimant workman is that of the insurer. The insurer's responsibilities are greater than those of all other parties to the claim, and the exposure to the possibility of error is also greatest for the insurer.

However, the insurer is also capable of reducing the risks of error in claims management that contribute to disability. This calls for a shift in the present-day emphasis in claims management if the insurer wishes to prevent disabilities.

The basic philosophy underlying current claims management appears mainly oriented toward compliance to workmen's compensation law. Disability prevention calls for perceptual claims management.

While it is not the intent of this text to detail the aspects of perceptual claims management, there is some need to clarify the differences resulting from the two methods. Management by compliance is self-defeating because:

1. The policy serves only to meet the minimums prescribed by workmen's compensation law;
2. A minimal response by an insurer furthers the adversary process;
3. The policy leads to internal practices and attitudes that perpetuate the anti-insurer mood of claimants.

Perceptual claims management is counter to management by compliance. The philosophy underlying this method of management is result oriented. Perceptual claims management causes:

1. The claims processor to sense the individuality of the claim and the nature of the specific person, or claimant;
2. The claimant workman to view the insurer as an organization of individuals concerned and responsive to his, or her, predicament;

3. The integration of the many benefits and services encouraging recovery;
4. The claims processor to view success by the values resulting from preventing a disability rather than by the dollars saved temporarily through a minimal response.

Aside from perceptual claims management so important to disability prevention, the insurers face other obligations. Many criticisms of insurers exist. The future of the private insurer's role in workmen's compensation is dependent upon the industry's ability to correct certain deficiencies.

The address of Paul S. Wise[16] contains six guidelines for corrective action. While all insurers will do well to study Wise's recommendations, one of these important guidelines needs further consideration here. This is research.

The depth of our ignorance stemming from the lack of meaningful research in workmen's compensation is appalling. Under scrutiny, the problems facing the claimant-workman and workmen's compensation agencies differ little from those at the turn of the century.

The ability to solve these problems lies in the wealth of knowledge buried in the claim files of insurers and workmen's compensation agencies. They contain the stuff needed to perfect the system.

For example, indications of abuses and substandard performances by many physicians exist. However, the insurers and the agencies are incapable of effecting corrections because they do not perform the research needed to take the matter before the medical professions.

Poor medical care combined with poor claims management is the surest way to disable a claimant-workman. Neither are permissible in disability prevention. Both are correctable through proper research by insurers.

Another aspect of workmen's compensation in which insurers have a responsible role is the second injury relief plan. The availability of work for the workman is a factor in disability prevention. A practical second injury relief fund encourages employers to hire the workman having an impairment.

Today insurers are suspected of discouraging employers i

hiring of handicapped workmen. A means of dousing these suspicions is for the insurers to take an active role in promoting
better second injury relief legislation.

In most states it may also take certain legislation to implement
a disability prevention plan. Although all insurers are capable
of implementing their own plans, disability prevention is better
a function of a neutral body such as the workmen's compensation agency.

If an insurer chooses to embark upon a disability prevention
plan, it is important that its function not be subordinate to
claims processing. It is not the purpose of the disability prevention personnel to serve as investigators for claims processing.

In fact, the flow of information should always be from claims
processing to disability prevention. Because it is absolutely necessary to build a feeling of trust between the workman and disability prevention personnel, all information developed by disability prevention must be treated as privileged.

Information provided by the disability prevention division to
claims processing should be limited to the workman's rehabilitation plans or the notice of return to employment.

Because this is an extremely sensitive area in disability prevention, it is in part a reason for insurers to encourage the
formation of a disability prevention program under a neutral
workmen's compensation agency.

The role of the insurer where the workmen's compensation
agency administers the disability prevention program is simpli-
fied to some degree. The insurer's chief functions, aside from
its usual function of claims management is to:

1. Make early selection of the workman in need of disability
    prevention service;
2. Make prompt referral of the workman to the disability
    prevention division;
3. Establish and maintain closer direct contact between the
    claims processor and the workman;
4. Cooperate fully with the recommendations of the disability
    prevention division.

An insurer's reaction to their first brush with the disability prevention concept may be adverse. The insurer may view the concept as mollycoddling the workman and a waste of money. The insurer may even view disability prevention as another name for rehabilitation. Disability prevention is neither, and when properly implemented and applied, it is a worthwhile act of prevention for the workman, the insurer, and the workmen's compensation system.

EMPLOYERS

All employers can learn something from the recently implemented Occupational Safety and Health Act. This act causes all employers to demonstrate their dedication to creating a safe workplace for workmen, by law, and with this, the employer is forced to live with the good and bad of that law.

But injuries still happen. Workmen are still disabled by their work injuries. Comes now an increasing concern over the disabled workman. Will the problems of disability and the disabled workman be solved by more law? Or, will employers take the lead in preventing disability in the workman?

An intense voluntary effort by employers to prevent disability may forestall the passage of a federal workmen's compensation act. Or, at least, a dedication to disability prevention may enable employers to bring about a federal workmen's compensation act that is more desirable from the employer's viewpoint.

Many of the errors committed by employers which contribute to an employee's disability are mentioned elsewhere in the text. It is not the intention in this section to discuss the detailed actions or steps employers may take to prevent disability.

Disability prevention begins when an employer recognizes that the employee represents a major investment. It makes sense to protect this investment because a disabled employee represents a loss.

A normal human employer-employee relationship in the event of an injury is a step in preventing greater loss. Even the slightest interest in the injured employee by an employer is an act of disability prevention. This is a minimal effort on the part of an employer, and although it has value, it is insufficient as a disability prevention effort required by employers.

Employers are fortunate because disability prevention is not solely their battle. The other parties to workmen's compensation are also equally responsible in creating an effective disability prevention plan. However, knowledgeable employers will exert their influence on the other parties to join in the effort.

The majority of employers insure through a state fund or private carrier. The insurer chosen by any employer is an agent of that employer. In effect, the insurer is a reflection of the employer's attitude toward employees.

There are insurers who practice the concepts of disability prevention. There are also insurers who function purely on the profit and loss basis with no regard to the outcome of an injured workman.

The insurer manifesting a primary interest in dollar values is often in conflict with the claimant-workman. Conflict, or stressing the workman, is a major cause of adversary proceedings. The adversary process never lessens the workman's disability whether he wins or loses during those proceedings.

Since the employer is a consumer of the insurer's product, the employer is in a position of selectivity. The employer is wise to purchase insurance from an insurer practicing the concepts of disability prevention. This is one major step an employer can take in building a team dedicated to disability prevention.

Putting a workman back to some form of work at the earliest possible moment is another positive step in preventing disability. The work need not be full-time to be effective. Nor should the work assigned ever exceed the workman's capacity during his recovery from illness or injury.

The benefits from the early return to work for the workman and employer far outweigh the criticism by the detractors. This is a concept in need of greater recognition. Here again the employer is in a position to exert influence on the other parties to join in the effort to return the workman to work sooner.

In any community one of the reasons for the delay in returning the workman to work is the physician's failure to release the workman for work.

Of the various reasons why a physician delays the release to work, one reason is a fault of the employer. The majority of employers fail to make the proper efforts to keep the physician informed. As a result, the physician in his ignorance prolongs the recovery of the workman to the detriment of both the workman and the employer.

The fear of liability by employers in second injury to a workman is real. However, it is correctable. It requires that employers join the other parties in creating effective second injury legislation.

Employers are increasingly faced with pressure to employ or re-employ handicapped workmen. Again, legislation is forcing the issue. An example is HB 3057 passed during the 1973 Oregon legislative session. In summary, it reads as follows:

HB 3057 (Effective October 5, 1973)

Makes it unlawful employment practice for an employer to discriminate against a workman with respect to hire or tenure or any term or condition of employment because a workman has applied for benefits or invoked or utilized procedures provided for in the Workmen's Compensation Law or the Occupational Disease Law. Requires reinstatement of a workman who has sustained a compensable injury to his former position or employment which is available and suitable upon demand if a workman is not disabled from performing duties of such position or employment. Requires reemployment of a workman who has sustained a compensable injury and is disabled from performing duties of his former regular employment at employment which is available and suitable upon demand; subjects such reemployment to provisions in valid collective bargaining agreement.

An earlier commitment by Oregon's employers to an effective disability prevention plan would have circumvented the need for such legislation. Now that HB 3057 is law in Oregon, the employers have an obvious reason to perfect the present disability prevention plan.

Employers are rightfully concerned with the increasing costs of workmen's compensation. Legislating laws do not, nor will not, reduce a workman's disability nor the costs involved.

Aside from disability income and medical care benefits payable to the workman, disability prevention reduces costs because it:

1. Reduces the necessity for more legislation;
2. Reduces the need for vocational retraining;
3. Reduces the need for job modification, if the disability is adequately prevented;
4. Reduces the incidence of litigation.

Employers have large responsibilities in the planning of an effective disability prevention program. The benefits available to the employer from the program are also large. Whether or not a disability prevention plan is implemented in any state is largely dependent upon the employers of that state.

A voluntary, effective disability prevention plan leaves employers alternatives. Legislation restricts the employers' alternatives. Which concept prevails in the future is presently the employer's choice.

GETTING IT, TOGETHER

The best planned disability prevention program is doomed to failure if the support team rejects the proposition. The prime members of the support team are organized labor, the legal profession, and legislators.

Although the legal profession has a more direct involvement in a workman's claim than the other two parties, the attorney's role presently is not one that makes him a prime member of the care team. Perhaps the attorney belongs there, but not as the matter stands now.

The support team members' roles are no less important than those of the care team in preventing disabled workmen. This prompts some need for discussion of their roles in this text.

Disability prevention requires that we get it, together. Failure to involve all parties to workmen's compensation makes for errors in concept, fragmentation of effort, limiting actions and an effective disability prevention program. Thus, there are some matters to consider.

Organized Labor's Role

The greatest benefit for any workman claimant to workmen's compensation benefits is that of full recovery and freedom from disability. A workman who maintains his employability is far better off than a workman with disability.

Disability prevention also calls for us to teach workmen the value of preventing disability in themselves. Organized labor is better suited to this task than all other parties in workmen's compensation. Organized labor is the choice here because the teachings then stem from the workman's own organization, from men of his own kind.

Organized labor's efforts to date are centered upon improving the workmen's compensation benefits and creating safer work places. Just as teaching workmen job safety is possible, it is also possible to teach the workman that disability prevention is sensible.

Consider the matter of the workman and his present disability benefits. Testimony before the National Commission indicates that many workmen are robbing fellow workmen of greater available benefits.

Time-loss benefits paid to a workman unnecessarily disabled lessen the benefits for a workman badly disabled. Although it is not always a workman's fault, there are workmen who claim disability and draw time-loss benefits merely because workmen's compensation insurance exists.

Impressing upon fellow workmen the effect unnecessary time-loss benefits paid to another workman has upon their own overall benefits is the best method of stopping the improper usage of workmen's compensation insurance.

The average workman is not seeking undue benefits from workmen's compensation. However, many workmen do not fully appreciate the impact a disability award has upon their employability until the experience is encountered.

Through the educational process many workmen can be taught the pitfalls of disability—the economic losses, the emotional losses, and the social isolation brought on by disability.

It is also possible to teach workmen to be more alert to impending disability and thus enable them to avoid an unnecessary disability. The physical limitations of some workmen when combined with their trade permanently disables them after a few years.

Aside from teaching the workman the detriment of disability versus the gains in maintaining one's employability, organized labor can greatly benefit its members by actively assuming a role in the vocational rehabilitation of the workmen.

Where else does a more knowledgeable, a more authoritative group of people exist who are capable of assessing the workman's potential vocational abilities than within the structure of organized labor?

Although the members of labor unions are not viewed as educated men, the bulk are intelligent men. Beyond this, they are often practical men, having a canny sense in evaluating another workman's abilities. Their skill and expertise in their own trade enables them to determine another workman's ability to learn their trade. Beyond this, the labor unions are in a better position to assess the needs in various trades or crafts. It does little good to train a man in welding, or barbering, or bookkeeping, when the present supply exceeds the demand.

Because the average workman is not one of the dreamers of our nation, the future has less meaning than the present to him. This type of man is more practical in a sense. He prefers getting on with a job.

For the workman of this nature, on-the-job training is the preferable method of vocationally rehabilitating him. The workman first of all finds himself amid fellow workmen actively engaged in the pursuit of their jobs and their lives. Under these circumstances, there is less likelihood the injured workman will meet those things which reinforce his own disability.

Two immediate benefits occur for the injured workman in using on-the-job training as a means of disability prevention. The first is the association with the non-disabled workman. The second is the fact that on-the-job training meets the needs of the workman to relate to the present.

This form of vocational rehabilitation offers immediate income during the training. It promotes the immediate possibility of job security. It also aids in restoring human dignity to the workman because he is regularly employed.

Although all forms of vocational retraining play a part in workmen's compensation, it is possible to circumvent much of this effort through job modification.

Organized labor has a role in encouraging job modification. Much of the current temporary total disability imposed upon workmen is avoidable by altering a job to fit the workman's limited abilities.

Greater usage of job modification for earlier reemployment of injured workmen is possible if labor unions will:

1. Make job modification a part of the contract.
2. Allow reasonable job shifting which gives the employer a greater latitude of job modification usage.
3. Conduct studies to determine reasonable methods of modifying the jobs of a specific employer under contract.

A major problem facing any workman having a permanent physical impairment after injury is re-employment. It does little good to perfect vocational rehabilitation if the workman is not rehired.

Organized labor has an obligation in the perfection of sensible second injury relief funds that encourage employers to hire handicapped workmen.

Although most states have second injury relief plans, not all are sensible. They do little to promote the hiring of impaired workmen. Some second injury plans provide relief to the employer only in the event the workman is permanently totally disabled.

The manner of funding second injury relief plans is of interest. All sorts of methods exist, but in general the insurers or employers are taxed to provide the money. Oregon's method of funding appears to be unique.

In Oregon each workman contributes five cents per working day which is distributed into three different funds in the following manner:

1. Retroactive reserve fund — 3¢ day
2. Second injury fund   — 1¢ day
3. Administrative fund   — 1¢ day

The retroactive reserve fund's purpose is to provide what is essentially cost of living income benefit increases for claimants on a periodic basis. The income benefit increase itself is not retroactive. The retroactive aspect speaks to the date of injury.

The one cent per day contribution into the second injury reserve fund is used to encourage employers to hire the handicapped workman. For the employer who qualifies for second injury relief, the money is paid directly to the employer and not to the insurer.

Lastly, one cent a day is placed in the administrative fund for the administration and acquisition of equipment and staffing of a rehabilitation facility. This money is providing the funds for the construction of a modern, full-service physical rehabilitation center in Oregon.

This method of funding gives the working people of Oregon a greater voice in income benefits and disability prevention. Organized labor in Oregon is a cooperative team member with industry in perfecting the state's workmen's compensation act.

Also, as demonstrated in Oregon, organized labor is capable of, and has a responsibility in, perfecting the workmen's compensation acts. As an example, Mr. Whalen, President of the AFL-CIO in Oregon, took a position of leadership in developing the state's disability prevention plan.

Because physical rehabilitation is an important aspect of disability prevention, labor and industry agreed that Oregon's workmen were entitled to physical rehabilitation of the same professional quality as surgery provided by the best surgeons under the best of conditions.

The aim in Oregon is to provide physical rehabilitation of the highest professional quality in a properly constructed center.

This attitude in Oregon is a result of the persistence of organized labor. No reasons exist why any workman should receive second-rate physical rehabilitation in an improvised hole in some hospital basement or other such inadequate facility.

When needed by a workman, physical rehabilitation is as important as surgery. Both are of vital importance to recovery. Less than the best in either is an act of negligence.

Lastly, organized labor and all workmen have a responsibility to the workmen's compensation system. It is organized labor's obligation to both inform workmen of their rights under the act and to teach workmen their responsibilities.

The fully informed workman is more apt to make a better recovery, and this is disability prevention. The uninformed workman dependent upon rumor and erroneous information is likely to see the workmen's compensation system in a bad light, and this contributes to disability.

The Lawyer's Role

The legal profession is responsible for much of the progress in workmen's compensation law. However, as a profession, the legal fraternity has a responsibility extending beyond the perpetuation of litigation.

In the main, the immediate benefits gained for the workman in the adversary process are at best self-limiting, and do not meet his ultimate need.

From one point of view, gaining an increase in the disability award for the workman through litigation is akin to an insufficient amputation for a gangrenous leg. Both worsen the person's situation.

It is not solely a lawyer's task to obtain a disability award for a workman that aggravates the workman's predicament. The legal profession has an obligation to also seek benefits due a workman that prevent or reduce his disability and thereby better his lot in life.

No workman benefits by a disability, not even a permanent total disability. While a disability award may help the workman to survive, it is still wrong if a better solution exists.

From one side the legal profession is praised for its herculean efforts to upgrade workmen's compensation. From the other, it is damned. The damning is the fault of the profession itself.

In its concern for the inequities in workmen's compensation

law, the legal profession allowed legislators to circumscribe the role of the attorney in a manner that restricts the profession. The profession now has an obligation to break out of this sacred role for the benefit of workmen.

Whenever a workman seeks an attorney, it is the attorney's advice the workman is likely to heed from there on. The best manner in which any attorney can help a workman is to guide that workman back into a normal life with the least possible disability and an opportunity for a gainful occupation.

To serve the workman in this manner, there is a serious need of the legal profession to create alternatives to litigation for attorneys in workmen's compensation. This does not mean the workman's rights of appeal are cast off. It means the attorney, in his professional wisdom, has an opportunity to guide the workman in any manner that is in the workman's best interest. More disability is seldom to the workman's best interest in an equitable system.

It is impractical, of course, to consider that attorneys will choose disability prevention for the client unless there is recompense for the time and effort expended by the attorney in guiding the client in this direction. Therefore, the attorney who guides his client through the course of disability prevention, be it physical rehabilitation, vocational rehabilitation, or whatever, so that in the end his client is gainfully employed, is entitled to a fee for services rendered.

Legislators' Role

The greatest weakness in workmen's compensation is its delivery system. In nearly all states many good benefits for the workman already exist but they are essentially denied the workman because administration of workmen's compensation laws needs revamping and firmer backing.

It is a function of legislative bodies in all states to assure the citizens of their state that workmen's compensation law serves as a living instrument.

Legislators should insist that the law, and what it provides, is used to its fullest extent in providing prescribed benefits for workmen. A law should never be used as a mechanical method of ruling or inhibiting purposeful action.

The states are faced with federal take-over. Why? According to *Accident Facts,* published by the National Safety Council,[3] work injuries now carry a 9.3 billion dollar tab in annual costs. A major factor in accounting for this enormous figure is the waste resulting from an uncoordinated delivery system in workmen's compensation.

The deficiencies in the delivery system are clearly a criticism of the National Commission in its report. What the Commission appears to be saying is that too many claims are processed rather than managed; medical care is left to run its independent course; rehabilitation is more a happening than a planned event; and lastly, excessive litigation and the delays caused by litigation fixes disability in the workman's mind, thereby disabling him more.

In part, the blame for the disabled workman is laid by the Commission at the door of the administration of workmen's compensation law. Passive administration appears prevalent in too many states because it is hamstrung by the legislatures. Lack of leadership and initiative stems from lack of proper funding and staffing within many agencies.

Until legislators insist upon active leadership in the administration of their workmen's compensation law, the waste is going to continue. Furthermore, patchwork amendments or even complete revamping of the law will increase the losses unless administrative expertise is first established.

The philosophy of any workmen's compensation administrative agency is a reflection of legislative attitudes. If workmen's compensation legislation occurs as a result of stand-offs and compromises, delivery of the prescribed benefits tends to follow in a similar manner.

As an example, take a situation in which a legislature agrees to raise temporary total income benefits for workmen in their state. The increase is gained only after agreements are reached that continue to permit insurers to use questionable compromise and release settlements in closing claims. In effect, this nullifies the legislative action.

Under these circumstances the insurer is perfectly capable of applying pressures upon the workman in a manner designed to bring the claim to a premature conclusion. The administering body is likely to condone the insurer's action because it appears to be the legislative intent. The increased benefit voted for the workman is then lost to the workman.

The National Commission notes some three hundred measures were enacted by the legislatures in 1971.[2] In spite of this action, the Commission feels many of the basic needs remain unfilled. It points to the fact that of the sixteen recommendations published by the Department of Labor, the average state meets only eight. There are ten states which meet only four or fewer of these recommendations.

Since release of the National Commission's report, there is evidence of greater activity by legislatures to move toward satisfaction of the Department of Labor's recommendations. However, enactment of these recommendations alone do not guarantee an equitable workmen's compensation system for a state.

Even in a state meeting only four of the recommendations, if each is delivered properly to the parties, this is a far better situation than a state meeting all the standards but having passive administration and a faulty delivery system.

In all states there are people having expertise in workmen's compensation. Many already are employed in the administrative agency. Others are employees of insurers. All see and know many of the faults in the system. Legislators will do well to seek the counsel of such people.

However, it is not merely a matter of using this expertise to enact patchwork legislation, or to rewrite a specific workmen's compensation act. First comes a sense of direction. The primary matter is for legislators to determine the objective of their workmen's compensation system.

With the objective identified, legislators must establish an administrative body dedicated to the objective. In setting up the kind of administration needed, the neutrality of its position is paramount to its ability to score the objective. Let the battle of self-interests between the parties swirl outside the confines of the administrative agency. Give it immunity from political influences.

If preventing disabled workmen is the objective of the agency, then effective delivery of the benefits prescribed within the workmen's compensation act is the logical thrust required of the agency. Under this intent, compliance then occurs as a natural event and is no longer the all-consuming action of the agency, which, as currently practiced, detracts from the effective delivery of benefits.

Presently, administrative agencies are incapable of moving in this direction until legislators find themselves willing to call for improvement in the delivery system. As recognized by the National Commission, few legislators have expertise in workmen's compensation. A basic need is for legislators to establish a sense of trust in their workmen's compensation agency.

Once this trust is established, legislators are obligated to back the agency in a manner exhibiting this trust. In return, legislators should rightfully expect the administrative agency to keep that trust.

Beyond an agency of people having expertise in workmen's compensation, legislators should expect initiative, ingenuity, and dedication of any administration that is properly funded and fully backed in its objective by a legislature.

In summary, it is legislators who must assume leadership in correcting the inequities in workmen's compensation and in assuring the proper delivery of the legislation they enact. Through intent and proper action, legislators are capable of effectively reducing the costs arising out of disabling injuries by establishing disability prevention as their aim.

Chapter Eight

DISABILITY PREVENTION: A MODEL

T HE INTENT OF THIS CHAPTER is to relate experiences, offer
suggestions, and provide some guidelines to those interested
in establishing a disability prevention program. It offers assistance, not gospel.

It stresses the degree of involvement and the commitments
that seem necessary. All states are capable of implementing a
disability prevention plan, but it is not a simple matter of a
decision to do so, nor is it accomplished by a snap of the fingers.

To emphasize this and to make clear the effort and planning
required, let's turn to the Oregon experience.

A REVEALING PROBE

In Oregon it began with the Workmen's Compensation Board,
which is the administrative body. Or better still, it began because of those people working within the agency.

Oregon's workmen's compensation law, and the administration of its act, is as good as those of most states and perhaps
better than those of many states. Oregon is fortunate to have
people, from the three commissioners heading the agency on
down through the administrative layers, who are filled with constructive discontent.

At the time, a certain pride existed in the fact that this agency
was one of a few in the nation owning and operating a rehabilitation center. It was also a general consensus that workmen
were being properly rehabilitated through the activities of this
center. However, one day a staff member asked the question,
"How are we doing in rehabilitating Oregon's workmen?"

So far as the agency knew, there was no major cause for concern. No external sources rocked the boat. No accusations were
being made. No call for a probe existed. But the question arose

and the agency set about to determine the facts. What its study revealed ultimately brought about a whole new concept in Oregon's workmen's compensation system.

The agency discovered that the operation of its rehabilitation center was, in fact, schizophrenic. The center posed as a place where they rehabilitated workmen. But in reality the agency found it to be nothing more than a place where workmen were examined and evaluated for purposes of claim closure.

Workmen enrolled in this center experienced no benefit by the treatment attempted. Of the workmen admitted 91 per cent were unimproved at the time of their discharge from the center. The study revealed three basic reasons for the poor results:

1. Late referral of the workman,
2. Too short a stay at the center,
3. An inadequate physical facility.

The average workman did not arrive at the center until 18 months after his injury. The average workman enrolled was 43 years old and had an average of a ninth grade education. Sixty-seven per cent had back injuries and 51 per cent had undergone one or more back surgeries prior to admission.

Furthermore, psychological examinations routinely performed upon all workmen at admission to the center revealed nearly all manifested significant psychopathology and were badly disabled.

What the Oregon Workmen's Compensation Board soon recognized was the fact that its rehabilitation center was acting as a dumping grounds for the treatment failures of the medical profession.

Compounding this problem was the condition of the rehabilitation center, itself. The agency now recognized that its center was a badly equipped, crowded, and depressing facility, poorly located, in the heart of a large city. It became apparent that this center was not conducive to the physical rehabilitation of men and women already dejected and sorely disabled.

They were discharging the workmen from the center after a mere 14 days of treatment. Any physical or occupational therapy undergone by the workman served more to evaluate his physical condition than it did to restore anything to him.

From this investigation, the agency came to the only conclusion possible. Its effort to physically rehabilitate workmen referred to the center was an expensive failure costing the system over one million dollars per biennium.

Turning to an examination of its record in vocational rehabilitation, the agency found it extremely difficult to obtain clear facts. In part, this difficulty stemmed from the fact that Oregon's Workmen's Compensation Board must contract for all vocational rehabilitation services with the Vocational Rehabilitation Division.

The Vocational Rehabilitation Division is a separate state agency under federal control in Oregon. Its statistics and statistical methods leave much to be desired when it comes to ordinary people attempting to gain a clear picture of input and output.

However, the Workmen's Compensation Board personnel compiled the following facts and published the following report.

VOCATIONAL REHABILITATION

An attempt is made here to develop a rational picture of our present vocational rehabilitation program. Vocational rehabilitation is included in this study because this should fall inside the concept of rehabilitation which is meant to be all inclusive.

We have had to base our facts and figures on the entire year of 1969. Thus some difficulties in correlation are experienced.

First we will examine the plans written by D.V.R. and approved by our vocational rehabilitation specialist. Some explanation of the plans and their costs are included.

I. *Plans Written 1969*

	Number	Cost	
Total original plans	323	$ 321,232.65	
Total supplemental plans	725	227,525.68	(1)
Total no-cost plans	147	000,000.00	(2)
Total cost of plans approved	1048	$ 548,758.33	
II. Total diagnostics approved	353	16,748,49	(3)
III. Administrative costs		216,000.00	
Total D.V.R. costs		$ 781,506.82	

(1) *Supplemental Plans*

The meaning of this type of plan is best explained by an example. We will assume a claimant has an original plan for him to attend a barber college. If his scissors should become lost or broken, they are replaced. This is considered a supplemental plan. Other examples for the need of a supplemental plan for this same man might be:

1. Need of an additional text book.
2. Claimant moves from Salem to Portland and there are transportation costs.
3. Claimant fails to pass his examination and requires more barber training.
4. Claimant requires a medical examination.

(2) *No-cost Plan*

These are plans written by the D.V.R. counselor for the express purpose of meeting a quota. In other words, the counselor will counsel a claimant in a guidance session and this is considered a plan.

(3) *Diagnostics*

One example of this is where a particular claimant is referred to Goodwill Industries for a try-out period of time to see if he can embark upon an approved plan.

Another example is where it seems desirable for the claimant to have a medical examination, or a psychological evaluation prior to plan approval.

In the year 1969 we find 234 workmen completing their original plans. These plans are categorized as follows:

I.	Clerical and Sales	54
II.	Repair and Maintenance	31
III.	Welders & Machinists	53
IV.	Barbers & Beauticians	12
V.	All others	84
	Total	234

Further examination of the facts available gives us somewhat of an idea of our probable results obtained and the costs:

1969

	Number	Percent
Number of cases closed	778	100%
Total number employed	241	31.0%
Total number employed in field of training	135	17.4%
Average cost per claimant employed	$3,242.77	
Average cost per claimant employed in field of training	$5,788.94	

The discrepancy between the total number of cases closed and the total number of original plans written for 1969 exists probably because some of the plans closed were written in the prior year or years.

If we look at our total D.V.R. costs of $781,506.82 and consider this against the total number of original plans written (32) we find an actual cost of $2,419.53 per original plan written in 1969.

Recommendation:

A deeper and more time-consuming inquiry into our involvement in vocational rehabilitation is required. From this minimal material presented, a thorough study of D.V.R. plans and training categories is very desirable.

At the present time, we appear to be locked into an uncontrollable dependency upon D.V.R. to reestablish our claimants as part of the State's work force.

There should be given serious consideration to the development of our own internally controlled job placement program.

This should be done if a study of other states' programs where the use of D.V.R. is minimized reveals such a plan can be more effective in re-employment of injured workmen.

WELL!

Nothing in this entire probe revealed anything complimentary to the administering of the law by Oregon's Workmen's Compensation Board. They saw their heretofore proud efforts to rehabilitate injured workmen as an expensive farce.

When the Workmen's Compensation Board took this critical look at itself in 1969, this agency itself was only two and one half years old. A complete revision of the workmen's compensation law in Oregon established the Board as the regulatory agency and it began its function in January 1966.

At the time the Board became functional, it inherited the entire rehabilitation program and the rehabilitation center from what was previously the State Industrial Accident Commission, a monopolistic state fund.

Well! With its findings before it, the Board gulped, blushed, hitched up its britches and decided this was no way to run a railroad or a rehabilitation center. It asked itself, what were the basic needs enabling it to correct this grave problem?

The Board concluded the basic need was an aim and it set down its aim as *disability prevention.*

How, then, could the Board accomplish its aim in Oregon? It began by simply writing down five basic needs as it saw them:

1. Early contact with the claimant-workman and an evaluation of the workman's needs;
2. Coordinated and properly timed physical and vocational rehabilitation when needed;
3. The availability to a workman of a full-service rehabilitation center when needed;
4. An organized effort to re-employ workmen on a timely basis;
5. Full support by the legislature and all other parties to a workman's claim of any plan developed by the Board.

Fair enough. But how do you put lofty intents into actions? You do it first by educating others to the existing problem.

EDUCATION

Oregon's board made a clean breast of all findings in its study to organized labor, to industry, to the medical profession, the legal profession, and legislators.

The entire situation was laid before such men as: Edward

Whalen, President of Oregon AFL-CIO; Karl Frederick, Employment Research Director of Associated Oregon Industries; Robert McCallister of Georgia-Pacific Corporation; Chet Diehl of Weyerhaeuser Company; Charles B. Gill, Manager of the State Accident Insurance Fund; William T. Waste of Industrial Indemnity Insurance Company, and others. Each agreed it was time for change. All agreed to assist in formulating a new program.

In addition, two legislators, Dr. Morris Crothers and Keith D. Skelton, took particular interest in what the Workmen's Compensation Board was saying and proposing. They ultimately made considerable efforts in assisting the Board in its requests to the legislature.

Meanwhile staff members from the agency made trips to other states and Canada to learn what other agencies were doing in rehabilitation. Because the best coordinated plan appeared to be that of British Columbia, a committee composed of those previously mentioned from labor and industry and one legislator spent a week in British Columbia studying its program.

At the completion of all this fact gathering, the staff of the Workmen's Compensation Board began designing a proposed disability prevention program.

A SINGLE ENTITY

At the time in Oregon, approximately 120 insurance carriers, fifty self-insured companies, and the State Fund provided workmen's compensation coverage in the state. Thus the first matter considered was how any program developed could be made to function in an effective manner.

Is it always better for the private sector to conduct such affairs? If so, how do you assure the workmen in a state that the insurer will follow the concept of disability prevention in the workman's behalf? One method is by a directive from the administrative agency directing all insurers to do so.

But does a directive by an agency assure that the private sector will function effectively? In all probability, this requires at least that the agency establish some method of policing. But if you find insurers not in compliance with policy, do you then penalize the insurer? It can be done.

Policy directives and forceful compliance by an agency are always possible, but does this get the job done? More likely, in this situation, you merely find yourself back to the same proposition that is seen so often in workmen's compensation—action after the fact. For example, too often rehabilitation begins after it is determined that nothing else is feasible.

Then there is the matter of a workman's attitude toward workmen's compensation insurers. Like it or not, the average workman probably views the workmen's compensation insurer akin to the other fellow's auto insurance company. Both appear as adversaries to him.

In spite of denials, when the other driver crumples in your fender with his car, you very likely anticipate some sort of hassle with his insurer—like, they want the cheesiest repair shop in town to repair your fender. Even on those rare occasions when the other fellow's insurance representative is cordial and tactful, you find you harbor suspicions that somehow you're going to get the shaft, in spite of his smile and pearly teeth.

With this, it then becomes a question of whether or not an insurer's representative can be as effective as can another person representing a more neutral body, such as a workmen's compensation board not directly involved in insuring workmen.

Staff members and the committee of advisors concluded that allowing each insurer to operate its own program would not be as effective. In their opinion, it would probably perpetuate the present fragmentary manner of benefit provision and rehabilitation. They believed there was a need for a single entity.

The single entity is not the Workmen's Compensation Board. Rather, it is the Disability Prevention Division established in Oregon as an autonomous unit within the structure of the Workmen's Compensation Board. Autonomy is a key factor in the Division's ability to function effectively.

ORGANIZATIONAL STRUCTURE

The organization of the Disability Prevention Division is relatively straightforward as seen in Chart 1. The division is headed by an administrator and assistant administrator and contains three subdivisions:

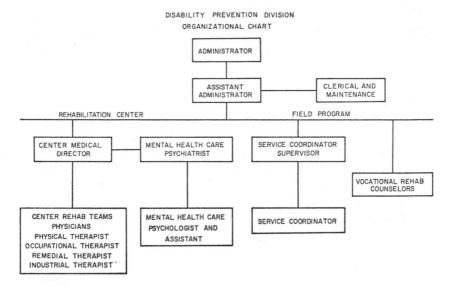

DISABILITY PREVENTION DIVISION
ORGANIZATIONAL CHART

1. Field Force
2. Rehabilitation
3. Clerical and maintenance.

The division is designed to perform three basic functions:

1. A field program
2. Re-employment
3. Rehabilitation.

Whose Program?

In designing any plan or program, the designers must carefully adhere to the aim before them. In this instance, it is disability prevention. But, whose disability prevention is in mind?

Obviously, it is the workmen's disability that needs to be prevented. It is not a case of preventing disability in the employer, the insurer, the physicians, or the agency.

Therefore the key function of the division is its *field program.* This field program must always keep the workman in sight. The best way to keep a workman in view is to be where the workman is, not where you want him to be for your convenience.

This is very basic to the reasoning behind the design of the field program in Oregon. Disability prevention begins where the workman is—on the farm, in the village or milltown, in the city or wherever. Working with, and within, the workman's environment is the clue to successful disability prevention.

Percentagewise, if the workman is contacted early enough and the proper evaluation of his needs and circumstance is made, it is seldom necessary to take the workman out of his own community. Because of this, the Oregon plan places its people operating as a field force in selected communities throughout the state.

This field force is composed mainly of people designated as *service coordinators*. Because the kind of person chosen as a service coordinator is so important, this aspect of the program is discussed in a separate chapter. Currently Oregon has service coordinators located in six of the larger communities in the state, and it is also planning to establish offices in other towns.

By strategic location of the service coordinators, the field program makes the services accessible to the workman and the workman accessible to the service coordinator. It also keeps the intent of disability prevention before the treating physician, the insurer, and the employer.

Because some workmen do need vocational rehabilitation, the field program also includes vocational rehabilitation counselors. The counselors are employees of the division and are not on loan from the Vocational Rehabilitation Division. They screen all workmen referred for vocational rehabilitation.

Following this screening, the workman is then referred to the counselors of the Vocational Rehabilitation Division for development of a plan. In every instance, though, the final plan is ultimately approved or disapproved by the counselors of the Disability Prevention Division.

Increasing use is made of on-the-job training for workmen in vocational rehabilitation. In the overall picture, less use is being made of V.R.D. because of the effectiveness of the service coordinators in getting disabled workmen re-employed through job placement.

Full Service Physical Restoration

A full-service physical rehabilitation center is under construction in Oregon for those workmen who need more physical rehabilitation than is available to them in their own communities.

This new center is the result of the planning of the Disability Prevention Division. The staffing and operation of this center is also a responsibility of the division. No other center now in existence for the physical rehabilitation of occupationally ill or injured in the United States will surpass this center and its capabilities.

Each workman entering the center will be assigned to a specific physician. Each physician is a member of a team composed of himself, a physical therapist, an occupational therapist, a remedial therapist, and an industrial therapist. It is the team's task to physically rehabilitate the assigned workman.

Upon entry the workman is thoroughly examined by the physician. A course of therapy to be provided by the other team members is prescribed. If indicated, the physician also requests a psychological evaluation of the workman. The physician may also request any other consultation he believes is important to rehabilitating the workman.

If psychological evaluation indicates the workman is in need of mental health care, this is initiated in conjunction with physical rehabilitative therapy.

A growing awareness in the importance of *mental health* in preventing disability of workmen caused Oregon to plan a mental health program for those workmen needing this care. The studies of Beals and Hickman [11] are partly responsible for this decision.

In addition, there are other indicators that appear to justify the establishment of a mental health care unit within the rehabilitation center. For instance, in a review of the permanent total disability awards issued during 1972, Oregon found that the psychopathology manifested by 101 workmen, out of a total of 204 workmen, was a significant factor in the decision to award them permanent total disability.

Furthermore, in another 634 workmen psychopathology was an influential factor in their unscheduled permanent partial disability awards. There is no indication in the records of these workmen that they received any mental health care in an effort to correct their problems.

Another growing concern in Oregon is about the number of

workmen using excessive amounts of prescribed pain-relieving drugs. Currently, studies are being conducted to determine as nearly as possible the severity of this problem.

With all things considered, Oregon's Workmen's Compensation Board believes there is a definite need for a planned mental health care program, headed by a psychiatrist. It sees the psychiatrist's functions as follows:

1. Evaluate and provide mental health care, when indicated, for any workman entering the rehabilitation center;
2. Assure proper follow-up mental health care through the treating physician and local mental health care units when necessary for a workman upon discharge from the center;
3. Develop a drug withdrawal program for workmen using excessive amounts of pain relief medication and mind-altering drugs;
4. Develop a behavior modification program for workmen manifesting aberrational degrees of pain;
5. Train the staff of the rehabilitation center in behavior modification techniques;
6. Conduct studies and research in mental health of workmen as it relates to occupational illness and injuries;
7. Enter into educational programs for industry, labor, insurers, and the agency.

The job set forth for the staff of the mental health care unit is a large one. However, it appears at this time that the workman in Oregon is not receiving the proper mental health care, when needed, to prevent his disability. Where indicated, a workman in Oregon is equally as entitled to proper mental health care under the workmen's compensation law as he is to the proper surgery. Therefore, the Disability Prevention Division is assuming a role of leadership to correct this present deficiency.

Oregon is placing considerable emphasis upon *industrial therapy* in the physical rehabilitation of the workman. This is not industrial training. Industrial therapy is work conditioning of a workman in a work atmosphere.

The type of work a workman will return to after recovery determines the areas of industrial therapy in which the workman is placed. For example, if a workman is likely to resume working in the timber industry, operating a chain saw in the woods, most of his industrial therapy will occur out-of-doors. He will be toughened into his job by operating a chain saw in a setting or rough terrain, with little heed paid to weather conditions.

The entire design of the industrial therapy department is akin to work areas found in industry. It is unlike the well-groomed medical atmosphere of the physical or occupational therapy departments. For this reason, the industrial therapy department is an industrial-like area physically removed from the main building of the rehabilitation center.

Emphasis upon industrial therapy is believed important to the restoration of a workman. When a workman completes a prescribed course of therapy at the center, knowing his level of workability is important. The workman's stamina and other work abilities are important to the employer hiring him. This knowledge is also useful factual information to those who must determine the final permanent disability award. But most of all, it is important to a workman to know his own physical capabilities.

Re-employment

Returning the disabled workman to a gainful occupation is the main thrust of the entire program. The chief deterrent to disability is meaningful and gainful work for the workman. Again, this is a prime function of the division's field force and it is discussed in greater detail later. It is sufficient to say here that Oregon is proving it can be done in a practical manner.

MORE EDUCATION

Implementing a disability prevention program in any state requires an understanding by many people of the need and the purpose. Failure to bring this understanding to the various parties normally involved in workmen's compensation can easily cause the program to fail if one is otherwise implemented.

As noted earlier in this chapter, the staff of the Oregon Workmen's Compensation Board began by bringing the need for a change to the attention of the major parties having key roles in workmen's compensation. However, the educational process did not stop here.

Both physicians and attorneys play major roles in the system. Educating treating physicians of Oregon to the needs of workmen began by meeting with committees of the Oregon Medical Association, the Oregon Osteopathic Association, and the Oregon Association of Chiropractic Physicians.

By bringing a full understanding of the existing problems and a plan for disability prevention to these members, they in turn served to gain the support of the full membership of their associations.

For example, at the 97th Annual Meeting of the Oregon Medical Association, Dr. Leland Cross placed a resolution before the House of Delegates supporting the Workmen's Compensation Board's disability prevention program. The House of Delegates adopted the resolution. A similar action also occurred with the Oregon Academy of Family Physicians.

Members of the Workmen's Compensation Board held informative meetings with attorneys active in representing both claimants and employers in workmen's compensation.

Members of the committee and staff of the Workmen's Compensation Board also gave informative talks before such meetings as The Occupational Safety and Health Conference, 1971, sponsored by the Chamber of Commerce in Portland, Oregon, the statewide membership of Union Shop Stewards and Business Agents, and the Oregon Self-Insurer's Association, which is an association of the larger companies in the state.

Such educational efforts are extremely important from at least two standpoints. First, it is a manner of gathering support which encourages the legislature to enact legislation necessary for effective disability prevention.

Secondly, education does away with misconceptions and squabbling due to misunderstanding that defeat the purpose. It is especially important that insurers, employers, and treating physicians fully understand the concept and the manner in which disability prevention functions because they are involved in the program on a day-to-day basis.

LEGISLATION

Very likely any state considering a disability prevention plan needs some legislative changes in its workmen's compensation act. What changes are necessary is dependent upon that state's present workmen's compensation law.

Two very likely items or measures that are needed from the legislature are permission to begin a new program, and funding for the program. It is a mistake for an agency to try sneaking in the back door with a piecemeal program that does not have the support and consent of legislators. However, one might consider that funding of a program by the legislature implies fully informed consent.

Through legislative action, funding in Oregon for the Disability Prevention Division comes from three sources:

1. The Administrative Fund, which is derived from an assessment of employers. This money pays the Disability Prevention Division administrator's salary, the salary of the secretary, the service coordinators' salaries and their additional costs such as office rentals, goods and services, and transportation.
2. The Rehabilitation Reserve, which is a one cent per day contribution paid by all workmen covered under workmen's compensation law in Oregon.
3. A direct charge to all insurers for services rendered to their insureds enrolled in the rehabilitation center.

The Oregon Workmen's Compensation Board's rehabilitation center is self-supporting in the sense that a fee for services is charged the insurers which covers the cost for its operation. The one cent a day paid by workmen is temporary funding for constructing and equipping their new rehabilitation facility. The employers' assessment funds the remainder of the disability prevention program.

Aside from funding, the next most useful legislation is second injury relief. The practicalities of this cannot go unrecognized. It is unreasonable to expect an employer to expose himself to the possibility of a permanent total disability claim against him because he hired a handicapped workman.

Oregon's second injury relief is not perfect, but it has been definitely improved by legislative action to assist the disability prevention program. In essence, for those employers qualifying for second injury relief, costs to the employer arising out of and attributable to the second injury are paid directly to the employer. Prior to the change in the law, such costs were paid as reimbursements to the employer's insurer.

Maintenance for workmen undergoing vocational rehabilitation is essential. In all probability, relatively few workmen really need vocational retraining, contrary to what we seem to believe. However, it is also unreasonable to evaluate a workman's potential, design a vocational retraining plan for him, and then cop-out on his bed and board.

At least in the state of Oregon, by the time a workman becomes involved in a vocational retraining plan, workmen's compensation has already invested a considerable amount of money in the workman's total care. For a few more dollars, it makes no sense to risk failure and a possibility of forcing the workman onto welfare because reasonable maintenance wasn't provided.

Oregon's legislature has corrected such a maintenance deficiency in its law. A workman undergoing vocational retraining is now receiving adequate maintenance because Oregon views this as economic wisdom.

CONSTRUCTIVE DISCONTENT

Once a disability prevention plan is designed and implemented, this is not the end. Constant monitoring, study, and evaluation of the plan must continue. It is the result, not the plan, that is the objective.

Any change in a plan arising out of factual studies should not be made with the intent of stopping something from happening; rather it should cause something to happen for the result. For example, let's reconsider our concern over our failures arising out of back surgeries which was discussed earlier.

There are surgeons ready to denounce spinal fusions because it appears that there are so many dismal results. In an effort to offset these poor results, there are people who are willing to set down dictatorial rules of procedure for all impending spinal fusions. Possibly they may be right.

One rule that has been suggested may be an excellent one,

but not as an intent of stopping a spinal fusion. The suggested rule is the mandatory requirement for all workmen to have a psychological evaluation before undergoing the back surgery.

This could be a very good rule because it causes something worthwhile to happen. Although the following may be objectionable phrasing, what is worthwhile behind this rule is simply this: if you take a kook into surgery, you come out of surgery with a kook.

In a study reported by A. M. White,[17] it was found that in workmen manifesting definite psychopathology, surgical results in 90 per cent were unsatisfactory.

A preoperative psychological examination causes those concerned to recognize in a workman manifesting psychopathology that which is disabling to him postoperatively. This then, should cause the workman to receive the proper mental health care before surgery which will give reasonable assurance that a successful operation will result. It does not stop the surgery. It merely causes a change in order to obtain a result.

Going on to other examples of constructive discontent, we might ask the question, "Is physical therapy a valuable treatment form for low back injuries?"

Proponents of physical therapy take a dim view of any one asking this question. After all, physical therapy is a tried and proven useful tool. Why else do doctors prescribe this treatment?

Perhaps the doctor's prescription is more business than it is therapeutic. In A. M. White's study,[18] he questions the efficacy of physiotherapy in low back injuries having pain of a discogenic origin. He also concluded that no form of physical therapy was beneficial to such a workman beyond six weeks and that the patient was worsened by continued therapy because it appears to aggravate any psychopathology which might be present in a workman.

Here again, such studies should not be used to stop a treatment form; rather, it should be used to evolve to a better treatment form yielding the desired result. Perhaps physical therapy is a useful treatment form in this case if combined with other specific procedures.

Constructive discontent should always combine with flexibility in a disability prevention program. As an example we can use the treatment programs in a rehabilitation center. It is probably a waste to routinely start all workmen entering the center in physiotherapy. It should depend upon what has happened prior to a workman's admission.

For example, in Oregon there are communities where the workman can receive the very best of physiotherapy. Thus, there may be no need to begin the workman on that program at the center. Each workman should be evaluated individually and the center should have the flexibility in its programing to start the workman where it will do the most for the patient.

The most serious threat to any program once it is implemented and working well is the complacency and rigidity of function that sets in. This is avoidable by maintaining a staff of curious people unafraid of getting egg on their faces at times.

In part, stagnation and ultimate failure is also avoidable by continuing education of both insiders and outsiders. Because it is easy to be blinded by one's closeness to a program, it is worthwhile to insiders to bring in qualified critics from outside for objective criticism.

There is one thing that must be rigidly adhered to in disability prevention. This is meeting the needs of the workman. That is accomplished by maintaining a state of initiative, inquisitiveness, and flexibility within the program.

RESULTS

Oregon's Disability Prevention Division and its program is still in its infancy. It is moving and it is proving itself. The results are beginning to stir the interest of heretofore uninterested people. The greatest gain from the Division's efforts presently is coming out of its field program.

This aspect of the program in Oregon began in January 1972 with the hiring of two inexperienced, untrained men. From then on until January 1973, these two men worked, learned, and developed the model that subsequent service coordinators will follow.

The results of the first year convinced the Division that it was on track and it began phasing in six more service coordinators.

A review of the results obtained by two service coordinators during the first year reveals the following:

Total referrals received	382
Total workmen re-employed	104

Breaking these facts down into more detail, they learned the following:

Total referrals received		382
Referrals rejected		−50
	Total	332
Acceptable referrals		332
Referred to V.R.D.		−28
	Total	304
Acceptable referrals		304
Closed for other reasons		−68
	Total	236
Employed old job		19
Employed modified job		41
Employed new employer		44
	Total Employed	104
Acceptable referrals		236
Employed		−104
	Total Open Cases	132

Job development is a major aspect of the service coordinator's job. In the first nine months these two service coordinators developed 180 jobs.

A question may arise concerning the apparent credit given for returning nineteen workmen to their old jobs. The service coordinator is not given credit for returning a workman to the former job. However, it is known in each instance that the service coordinator influenced the workman's return to his old job earlier than it would have occurred otherwise.

In the fifty rejected referrals, they found that these workmen

were already involved in other training programs, or that the
workman was not covered by workmen's compensation, or that
the workman was located too remotely from the limited areas
in which these two service coordinators functioned at the time
and it was therefore totally impractical because of the distances
involved.

Turning to the sixty-eight cases closed for other reasons, these
included:

7 who received social security or a permanent total disability
   award

7 who were already involved with the Vocational Rehabilitation Division

7 who found work without assistance

2 who left the labor market

3 who returned to school on their own

3 whose claims were rejected

8 closed for miscellaneous reasons

8 who would not cooperate in re-employment efforts

23 who moved or could not be located

Inquiry into the 132 open status cases revealed eighty-four
workmen were not yet released for work by their treating physician. The remaining forty-eight workmen were at various stages
of entry into the program.

Other factors were also reviewed by a sampling of these workmen's records:

Workman's average age		36.8 years
	Range	23-60 years
Average time from date of injury to referral		7.65 months
	Range	2.5-16 months
Average time from date of injury to employment		8.5 months

	Range	3.5-17.5 months
Average time from date of referral to employment		4 weeks
	Range	3 days-8 weeks
Average wage prior to injury		$3.66 per hour
Average wage at re-employment		$3.22 per hour

The most discouraging aspect of these facts is the delay seen in referring the workman to a service coordinator after the workman's injury. The delay is far too long when we consider that we are talking about disability prevention and not rehabilitation. However, we must keep in mind the total newness of this entire concept to the physicians, the insurers, and employers in Oregon during the first year.

A very positive and encouraging factor which emerged in the first year is the fact that the workman is being reemployed in a gainful occupation. A wage differential of twenty-two cents per hour is very acceptable when we consider that one-half of these workmen were beginning employment with a new employer.

Moving on to the second year of operation, between January 1973 and May 1973, six more men were hired, trained, and put to work as service coordinators. The last three coordinators did not begin working with workmen until May 1973. This is an important consideration in reviewing the 1973 results.

All service coordinators are assigned a quota of workmen they are to return to work each month. For the first three months, their quota is three workmen per month. By then their experience in the field is such that they are expected to place six workmen per month in jobs meeting the criteria of gainful occupation.

The following statistics are those for the eight months from January 1, 1973 to September 1, 1973:

Total referrals received	577
Total workmen re-employed	316
Total cases rejected	7
Total cases referred to V.R.D.	67

Employed old job with assistance   51
Employed modified job   174
Employed new employer   91

          Total employed   316
          Jobs Developed   719
Average wage prior to injury $4.03 per hour
Average wage modified job $4.23 per hour
Average wage new job $3.28 per hour
Average time from date of injury
   to referral   5 months
Average time from date of release
   to back to work   20 days

Although the delay from the date of injury to the time of referral during 1973 has been shortened by 2.65 months, it is still too long a delay. There is growing confidence that this time lag will be shortened in the coming year because of increasing familiarity with the program.

An encouraging aspect is the ages of those re-employed. Forty per cent of the workmen range in ages from forty-six years to over sixty-five years. There were actually twelve workmen over the age of sixty-five years re-employed.

In those workmen referred to the Vocational Rehabilitation Division, increasing stress is placed upon on-the-job training (O.J.T.). The length of O.J.T. depends upon the individual workman's situation.

Prior to and at the advent of the disability prevention program, V.R.D. took an average of six weeks time to develop an O.J.T. plan for the workman. Through the efforts of the disability prevention division, V.R.D. is now completing O.J.T. plans on an average of one and one half weeks.

THE SUM OF IT

In spite of many criticisms of workmen's compensation, the system is filled with many good benefits for workmen. The weakest aspect of workmen's compensation lies in the delivery and utilization of workmen's benefits. In this, too many states are remiss.

Each state has the ability to formulate a disability prevention plan assuring better delivery and sensible utilization of workmen's benefits. Presently, the least number of workmen get most of the benefits. Continuance of our present methods merely assures the least even more benefits with each attempt to improve benefits. Disability prevention brings the benefits to all workmen on the basis of actual need and thereby broadens the benefits in an equitable manner.

Before beginning a disability prevention plan, each state must objectively assess its weaknesses in its own system. This takes people of integrity, willing to face embarrassments over the facts they find because they must openly reveal their findings to other people.

No room for partisanship exists in disability prevention, nor among the people who plan such a program for their state. This takes considerable education of the people doing the planning because each likely represents a faction who is a party to workmen's compensation. Each group is used to jockeying for position when working out the compromises in workmen's compensation which result in more benefits for workmen.

However, in developing a disability prevention plan, these people must keep a clear mind. They are not hammering out compromises over benefits already legislated. Instead, they are putting to good use that which exists. They must keep in mind that good utilization and good delivery of existing benefits is a means to lessening the need for escalating benefits in the future.

The best plan drawn is one of simplicity designed for the convenience of the workman, not for the convenience and the benefit of those who implement the plan.

The plan is first and then the legislation. Careful evaluation and legal opinions on the existing workmen's compensation law may even preclude any necessity for legislative action in a state.

Day-to-day operation of the plan is best performed by a body that is administratively neutral. A high degree of flexibility should exist in the plan to assure that it continues to meet the needs and changing needs of workmen. This is done by critical evaluation of results.

It seems that workmen's compensation is at a fork in the road.

The assumption of workmen's compensation as a function of the Federal Government appears at hand. One of the arguments for federal involvement is the standardization of benefits and the delivery of those benefits in an equitable manner.

The states still have an opportunity to salvage their respective workmen's compensation systems, if they act. Immediate development and implementation of an effective disability prevention plan is a means of stopping federal intervention because it provides the ultimate benefit to the workman which even the Federal Government can't surpass. Such a plan makes an honest effort to prevent the disability in the workman. Disability prevention offers the workman an alternative to disability and dependency.

Chapter Nine

THE SERVICE COORDINATOR

THE KEY

THEORY AND PLANNING ARE IMPORTANT but unless they lead to a result, they become dreams and hopes. Because of this, understanding the concept of the service coordinator makes this probably the most important chapter in this text.

Although disability prevention began as a thought, a concept, what is written here is based upon experience in British Columbia and Oregon. None of it represents a sudden flash of brilliancy. It is a result of learning from past experiences, and of trial and error.

Call those persons working as service coordinators what you will, but failure to understand the kind of person needed for this job, or failure to give proper weight to the service coordinator's functions is the greatest error anyone can commit in planning a disability prevention program.

As for the name, there may be some merit in carefully choosing a job title for these people. Such names as Rehabilitation Counselor, Vocational Counselor, Employment Specialist, and such are pretty well shopworn. In fact, there is almost a stigma attached to such job titles today.

Ask any workman what image is brought to his mind by the word "counselor." We did in Oregon. The responses we received were mostly negative. To most workmen questioned, a counselor is someone sitting behind a desk, talking big talk, and offering damned little practical assistance. Whether or not this is true is not the point. The point is that the image is there in many workmen's minds.

Therefore, Oregon decided to call these people "service coordinators." It is a good title because they are just that—coordinators of services. Although such a job title may be foreign to a workman and to others, it has merit because it doesn't bring out a negative image in a workman's mind which may create a negative attitude from the onset.

However, the job title is not the key to a disability prevention program. The key is the person who is the service coordinator and the manner in which this person functions.

The service coordinator is why a disability program is successful. The service coordinator is not the sole reason why a workman recovers with little or no difficulty. The care team makes this possible. The service coordinator is the binding agent which makes the care team's action whole and effective.

NO EGGHEADS, PLEASE

Because a service coordinator's role is so important, considerable care needs to be exercised in the hiring of each one. Oregon recognizes this and puts each candidate through three screening interviews before hiring.

Program planners often look to the educated as their key people in programs of this nature. This is not the case in Oregon, however. The important people in Oregon's disability prevention program are essentially uneducated workmen hired to work as service coordinators.

Although they are little-educated people, each is an intelligent person. They spearhead the program, and their efforts are supported by others who are more educated such as administrators, physicians, therapists, employers, claims managers, and legislators.

Oregon hires its service coordinators primarily on the basis of the person's work experience and intelligence, not the level of formal education achieved.

This may, in a general sense, appear as a put-down of people educated in our colleges as social workers and counselors. It is not. It is simply a matter of proper utilization of people. Those educated in those fields are better utilized in the rehabilitation process.

The service coordinator is not involved in the actual act of rehabilitation. Therefore, work experience serves the coordinator better. Since it is the more educated who are most often the planners of such programs, it is their obligation to keep this matter of proper utilization of people in view. You don't send a general to the front to take part in a hand-to-hand skirmish. You send the man trained and conditioned for that purpose.

A well-educated person, steeped in sociology and psychology, is likely to be a washout as a service coordinator. Why? The reason, in part, is a matter of relationships. Consider the basic matter of communication between two individuals. The uneducated, or less articulate person, is often uneasy in association with a well-educated person.

As a result, the well-educated person does not make himself clear to the other person. For example, many doctors believe they make themselves clear in talking with their patients. However, there are strong indications that most patients do not understand their physician and that they go along with his care mostly on the basis of trust in his intelligence.

It is difficult to explain this problem of intercommunication between two persons of markedly different educational levels. Although the educated one tries hard to make himself or herself clear to the other, it just doesn't come across.

Perhaps it is explainable on the basis of the lesser educated person's apprehension. Maybe it is this which detracts or interferes with their ability to concentrate upon what the other is saying. Or is it on a basis of something no one fully understands at our present level of knowledge?

We may one day come to know that the silent communication between two people is more the basis of their understanding each other rather than the actual words they utter. Then again it may not be true. It may be the inflections of voice or a manner of pairing words. But whatever, one needs to spend little time with workingmen who are limited in their education to realize that they have their own way of understanding each other.

Workmen needing services provided by disability prevention are rarely well-educated people. Although a more educated claimant may be in need of physical rehabilitation, the well-educated claimant seldom needs vocational rehabilitation or any great degree of assistance in finding employment after recovery.

In fact, under close scrutiny we might find that the well-educated claimant automatically receives better care in the areas of rehabilitation than do the lesser educated claimants.

In examining claims of workmen in difficulty, and those often manifesting marked disability, we see that their education is commonly limited to the eighth grade, give or take a couple of years. The bulk also fall in the age groups between thirty-five and fifty years.

Ordinarily, these same workmen are not highly skilled tradesmen. To send a well-educated person to help such a workman often confuses this workman or even scares the hell out of him. Two G.I.'s make it better together on a battlefield than a colonel and a G.I.

Send as a service coordinator, a veteran scarred in the battles of the ordinary work world. A service coordinator trained by such experiences will make it better with the workman needing help.

In part, this is so because they understand each other. An understanding of the person you are trying to guide or help is of primary importance in disability prevention. For example, there are workmen who have had only one boss during their entire work lives. Suddenly being faced with the proposition of finding another employer may well overwhelm this workman. He doesn't know how to find a job, much less where, perhaps.

A service coordinator with a great deal of work experience in differing work situations knows how and where you get another job. It is likely he can also sense the fear and futility in a workman he is assisting because this service coordinator has also felt those same emotions in the past when he had to seek a job.

Consider the matter of wage loss after injury and the financial jeopardy a workman may suddenly face along with the despair arising out of such a plight. A service coordinator whose past experiences read like a job-hopper's credentials likely knows what it is to face a loss of regular income. By this experience the coordinator is more likely to sense any depression in a workman than is a treating physician or a claims reviewer.

Work experience gives the service coordinator abilities not attainable through formal education which is a mandatory qualification in disability prevention. The nature of his life's experiences gives a service coordinator the ability to identify with workmen needing the help found in disability prevention.

Beyond work experience and a good brain, there are other things to consider in hiring a person as a service coordinator. This person needs a fairly high degree of self-confidence, a definite liking for people and a desire to work with other workmen, and each must be the kind of person who is a self-starter.

In summary, each service coordinator must be an intelligent, practical person. The correctness of his English is not so important as his ability to make himself understood. He needs to be an easy talker who by life's experiences understands the ways of workmen and the problems workmen face. Leave the theory and the planning to the eggheads.

SELECTED EXAMPLES

Before hiring, each person in Oregon applying for a job as a service coordinator is first screened by Civil Service personnel. Those selected are then interviewed by a committee composed of a high-level organized labor leader, a representative of industry who is well-informed on workmen's compensation affairs, and an interested treating physician. The third and final interview of the applicant is by the supervisor of service coordinators who makes the decision to hire.

At each interview level the applicant is rated as first choice, second choice, and so on down. Although the supervisor is not bound to always hire the first choice applicant, it does serve as a guide for the supervisor.

This may appear as a troublesome way to hire service coordinators. However, the people supporting Oregon's disability prevention program recognize the vital importance of any person hired as a service coordinator.

Aside from the fact that disability prevention is a joint effort of the parties to workmen's compensation in Oregon, the three-member screening committee is picked for a purpose.

The labor leader is looking at the applicant's work experiences and abilities that relate to workmen. The industry leader is looking at the applicant's ability to talk with and sell employers on the concept. The physician is evaluating the applicant's emotional stability and ability to work with treating physicians. All are evaluating the applicant's native intelligence.

Although a great deal of stress is put into these writings to impress upon the reader that the level of a person's formal education is not critical to the hiring of him, there is bound to be a mixture of educational levels among service coordinators in a well-conceived disability prevention program.

To date, Oregon has hired nine service coordinators. Six have a high school graduate level of education. Two have attended two years of community college and one service coordinator has four years of college and a bachelor of science degree.

It is estimated that Oregon will employ twenty to twenty-five service coordinators when its program is fully implemented. Beyond an educational mix there is a necessity for a mix of sex and race. The ratios of these latter two factors is something that is a variable need from state to state.

The hiring of women as service coordinators in a given state should not be dictated by the male-female ratio of workmen in that state. The workman's sex does not determine the success of the service coordinator's efforts. The male service coordinator may be more successful with preventing disability in a specific female workman than a female service coordinator, just as the opposite circumstances may occur.

There is probably more reason to give attention to ethnic differences whenever a state is populated with large numbers of distinct ethnic groups. There is also cause to heed regional differences of the general population in a state.

Oregon is a state having greater regional differences in the general population than it does racial differences. Excluding the metropolitan area of Portland, where many of the black people are banded together in a section of the city, the other races of Oregon are well disseminated.

Distinct regional differences do exist in Oregon although its population is small compared to that of many states. Oregon is divisible roughly into three major regions of difference in its population—the coastal region, the mid-region, and the eastern region.

The coastal areas of Oregon are populated mostly by people in the forest products industry, fishermen, some who farm, and retirees or vacationers. In the mid-region are found most of the business people, factory and plant workers, large farming operations, a greater mix of races, and a population generally exposed to greater national and international influences. The people of the mid-region might generally be described as those people who are middle America.

Eastern Oregon is a region of vast space and thin population. It contains people of strong bodies, strong minds, and strong loyalties. Eastern Oregon is primarily a region of ranchers, ranch hands, and forest products people. Although some similarity of occupations exists, those of Eastern Oregon are a breed apart from those of mid and coastal Oregon.

Because of these differences, service coordinators in Oregon are also selected from, and employed in, the locales from which they stem. It would be an error to hire a man or a woman who has spent most of their working life in Portland, Oregon to work as a service coordinator in Eastern Oregon. Too much time would be spent by this person in learning the nature of the Eastern Oregonian, of relating to them, of learning the nature of the region's work places and jobs and job language.

In planning a disability prevention program, the planners must know and evaluate the people of their state if they hope to hire service coordinators who are effective. This may seem picayune. However, the ills of disability are nationwide, and it is a major concern. To prevent and eradicate disability requires the same rigid attention to detail we see in other preventive measures such as the prevention and elimination of poliomyelitis.

Let's turn now to specific examples of people Oregon hires to work as service coordinators.

The first man is a forty-eight year old male, high school graduate. He served over four years in the U.S. Navy as a ship's cook. After his discharge from the service he was employed twelve years in a pulp and paper mill where he held several jobs including the training of other workmen.

He left the pulp mill and worked for a highway paving firm for eleven years until he suffered a serious back injury. He remained disabled for the next two years, drawing workmen's compensation benefits. During this time he underwent considerable medical care and numerous hospitalizations.

Finally, after two years of total disability, the parties to this man's claim decided he should be vocationally rehabilitated. Because he expressed a strong interest in working with people, a vocational plan was developed to train him as a job developer.

He attended a community college for the next two years, taking courses designed to qualify him for employment as a job developer for the Vocational Rehabilitation Division.

Upon completion, the Vocational Rehabilitation Division expressed interest in hiring him, but they could not because he lacked the experience qualifications specified for the job.

However, the Vocational Rehabilitation Division did put him to work as a volunteer job developer for the handicapped. Under this arrangement, he received his automobile travel expense and a lunch allowance. But no pay.

The objective of this arrangement with the Vocational Rehabilitation Division was to give him experience so he could qualify for a job as a full-time employee. He worked as a volunteer developer without pay for one year.

After that year he was employed by the Vocational Rehabilitation Division until he was hired as a service coordinator several months later. This man is unquestionably an intelligent person who is a real self-starter. After one year as a service coordinator he was promoted to statewide supervisor of the service coordinators.

As supervisor he has built a team of service coordinators who are highly motivated and they are obtaining far better results than were anticipated by the program planners. This man's personal experience with a work injury and the workmen's compensation system is a motivating factor for him which he in turn is able to instill into others.

An example of another person hired as a service coordinator is a thirty-six year old male, high school graduate. He began working in a plywood mill after high school. During his three years in this mill he was promoted to day shift foreman of the general plywood operation involving all phases of plywood manufacturing.

He quit his job in the plywood mill because he wanted sales and bookkeeping experience. He was employed as an assistant manager in charge of sales and bookkeeping by a furniture firm for two years. He then took a job as a glazier with a paint and glass company where he became a journeyman glazier and installation supervisor.

After four years as a glazier he went to work as a machinist and became a journeyman machinist. Several years later he quit this job and became a car salesman until he was employed by the agency as a service coordinator.

In addition to his varied work life, this man is also a gunsmith and has served as a staff member of vocational training in a local high school during the evening hours.

This man was born and raised in a logging-lumbering community. He relates well to workmen and he is acquainted with many employers in his locale. He is obviously a self-starter, and he aims for the top in any job he takes on. To date he is one of the most effective service coordinators in the agency.

A third man hired as a service coordinator is an Eastern Oregonian having a long, detailed work history beginning after his graduation from high school. During World War II he worked at several jobs in a shipbuilding yard.

After the war he returned to Eastern Oregon and worked part time as a ranch hand. During this period he followed and performed in the rodeo circuit and hunted wild horse herds for his living.

Later he turned to transcontinental truck driving and then mechanics. For the next twenty-three years his work ranged from that of a mechanic to heavy construction and operation of many machines. He also worked as a millwright several years and a shop foreman on several occasions.

This man even tried his hand as a salesman selling new and used cars, and at one time sold life insurance. For two years prior to his employment as a service coordinator he was self-employed in farming and cattle ranching.

His life is essentially that of an Eastern Oregonian and fairly typical of many workmen in that region of the state. Because he knows the ways of the Eastern Oregonian and has had vast and varied work experience, he is very effective as a service coordinator working in Eastern Oregon.

THE JOB

The aim of each service coordinator is to return the workman to a gainful occupation at the earliest possible moment, and with the least possible disability.

It is obvious from this aim that the service coordinators' activities are based upon a return to work theme. Vocational training is secondary to gainful employment in their efforts. But, whatever effort is made in returning the workman to work, it is done by cooperation and coordination of all parties to the claim.

To accomplish the aim and intent of their jobs, the service coordinator's functions are seen as five basic ones:

1. Early contact with claimant workmen;
2. Appraisal of each claimant's needs;
3. Effecting an overlapping of the employment-rehabilitation processes, whichever they may be, for each claimant;
4. Holding together the physician, the claimant, and the employer triad;
5. Maintaining a degree of flexibility throughout the entire program in a manner to attain the aim.

Early Contact

It is impossible to contact a claimant workman too soon. Early contact is a key factor in preventing disability in the ill or injured workman. The earliness of contacting a claimant can mean the difference between gainful re-employment for the workman and the need to vocationally rehabilitate that workman.

A psychological advantage frequently exists with both the workman and the employer which can obviate the need later to vocationally rehabilitate the workman if the opportunity is grasped soon after injury. More employers are willing to modify a job for an injured workman a week after an accident than they are six weeks later.

Therefore early contact with the claimant workman also puts the service coordinator in early contact with a workman's employer. The workman is still fresh in the employer's mind, and it is more likely a sense of obligation is still in the employer's mind. Sometimes it is amazingly easy for the service coordinator to get a job commitment from the workman's employer if contacted early.

The mere fact his employer makes a job commitment is a real psychological boost to most workmen. This is valuable because maintaining a workman's morale is an important aspect of disability prevention.

No perfect single source of referral exists which enables the service coordinator to make early contact. It is a matter of actively seeking out claimant workmen in need of help.

The best sources of referral come from the insurers, the treating physicians, and employers. Even here it is a struggle to keep before those parties the importance of early referral.

It becomes a process of indoctrinating the insurers into the concept of disability prevention and early referral by the administrative and supervisory personnel of a disability prevention division. Since the insurer is the first to receive a notice for injury or illness and the medical reports, much rests upon the judgment of the claims handler in making early referral.

Since so much rests upon the actions of the person managing the insurer's claim, considerable effort is made on a continuing basis in Oregon to educate insurers to the concept and role of disability prevention.

A very good referral source is the treating physician. Here again, treating physicians need indoctrination. The physician is the one who can make the earliest of referrals. It is a matter of the physician knowing who the service coordinators are in the locale and the manner in which they function.

For this reason, in Oregon the Medical Director of the Workmen's Compensation Board conducts a personal introductory meeting each time a new service coordinator is assigned to an area. Individual appointments are made with the physicians. In some areas where the number of physicians is small, each one is contacted. In an area where there is a large number of physicians, a limited number of key physicians are called.

By appointment the medical director and the service coordinator call upon the physician. A personal introduction is made. A brief explanation of the program is given. In addition, a short written description of the program is left with the physician and key employees. It is a short but informative visit.

Lastly, the physician is asked for a commitment to refer workmen he treats early in the course of care to the service coordinator. In one and one half years, not one physician has rejected the idea of disability prevention. All have committed themselves to early referral.

Employers are also a source of early referral. Again, it takes education. An interesting thing is occurring in Oregon among employers as they become familiar with the disability prevention program. More employers, including some large self-insured companies, are beginning to see the benefits. These employers are referring injured employees to service coordinators even though they intend to re-employ the workman.

What employers in Oregon are learning is the value of proper guidance of a workman through the workmen's compensation system and the value in this to both the workman and themselves. They see the worthiness in the ability of a service coordinator to maintain a workman's interest in returning to work.

A daily reading of local newspapers is another means of finding claimant workmen likely in need of disability prevention services. Most local newspapers report serious occupational injuries and it is a simple matter for the coordinator to make contact.

Of course, regular and frequent checks with local hospitals is another way of contacting workmen who have a potential need for assistance. Sometimes labor unions, shop stewards and such, make the referral because they have learned of the program through informative meetings. In the final analysis, a service coordinator must seek, and expect to receive, a referral from any source. The important thing is to find the workman.

Procedure

Appraisal of a claimant's needs is the first matter once a referral is made. Each claimant is an individual, as are the needs of each.

Upon receipt of referral there are three immediate steps the service coordinator takes in every instance. These are:

1. Contact with the treating physician;
2. Contact with the employer;
3. Visit with the workman.

Although these steps do not absolutely need to be taken in this order, the service coordinators in Oregon find that taking these steps in order works out best. The treating physician's office is contacted either by telephone or by a personal visit to the office.

Being in touch with the treating physician serves a twofold purpose. First, it is a matter of care team cooperation. The treating physician should always be aware of the service coordinator's involvement. It is also important that the service coordinator obtain the treating physician's permission to become involved.

Secondly, the treating physician's first report, or subsequent reports, do not always reveal all the pertinent medical facts. Therefore a service coordinator can obtain any additional information at this time.

A point to make clear about these contacts with treating physicians is the desirability of minimizing the physician's effort and conserving his time. Service coordinators in Oregon find it is rarely necessary to meet the physician face-to-face. The physician's permission and the necessary medical information is obtained through the physician's office personnel in the majority of instances.

Oregon has made a special effort to avoid any increased demand upon the treating physician. Therefore, they do not require any special reporting efforts on the part of the physician in this disability prevention program.

Periodic reports are made to the treating physician by speed letters from the service coordinator. The physician is always informed immediately when the service coordinator obtains a job commitment from an employer. This report describes the job, the job location, and the wage the workman will earn.

Once the job commitment is obtained, it is left to the treating physician to determine when the workman can resume work. If it is not immediately, the service coordinator keeps in touch with the physician until the workman is released for work.

Occasionally a treating physician finds that a service coordinator is of additional help. If a physician requests it, the service coordinator will report the workman's home and living conditions to the physician. Such information is at times invaluable to a treating physician.

The relationship between the treating physician and the service coordinator has an additional effect which is not as obvious. It brings about a greater awareness in the physician of the person undergoing treatment. As a result the treating physician consciously is more aware of the passage of time, of the length of treatment, of the time loss involved, and of the workman's response to treatment.

This effect upon the physician is a subtle one created by association. It is not the result of a tangible reminder. Never is any attempt made to openly question the physician about the treatment or length of care. All service coordinators are clearly instructed that they are never to question any physician's professional care.

The workman's employer at the time of the injury or illness is the next person contacted by the service coordinator. The employer is simply asked if the workman will be re-employed when released for work by the treating physician.

Underlying this is the belief that for most workmen it is easier to resume working for the same employer. There are fewer adjustments for the workman to make.

However, the employer may feel that there is no way to re-employ the workman because of the nature of the work and the nature of the workman's injuries. This is where a service coordinator's work experiences and ingenuity are invaluable.

With the employer's permission, the service coordinator tours the work place. Often there are jobs suited to modification which a service coordinator having good work experience will note. Of course, job modification must be applicable on a sensible and a practical level.

Many employers will agree to modify a job for a workman when a reasonable manner of doing so is brought to the employer's attention by a service coordinator.

In addition to this, a service coordinator also learns something of the workman from the employer and of the employer's prior relationship with the workman.

The last and vital step is a visit with the claimant workman. This is an informative visit and is not an interview with the workman. Information usually passes both ways in this first visit.

The service coordinator goes to the workman's home, or if the workman is hospitalized, to the hospital. In the instance of a hospitalized workman, the service coordinator always makes at least one visit to the workman's home later when the workman is released from the hospital.

Insistence upon these home visits is made with a purpose. A service coordinator can gain much insight into a workman as a person by a home visit. An idea of a workman's pride in himself and his family is often readily apparent. The workman's financial plight may be very visible by his living conditions. If a workman has a wife and children, their attitudes toward the workman and the total situation are noted when meeting a workman's family. Many things are learned in addition to these examples.

From the other side, this visit is intended to be informative for the workman. The first thing a service coordinator makes certain of is that the workman has a clear understanding of the rights he or she is entitled to under workmen's compensation law.

Because many workmen do not clearly understand the benefits coming to them, or have a faulty understanding, this is an aspect of a service coordinator's function which cannot be overstressed.

Of course the service coordinator makes some appraisal of the workman's attitude toward returning to work and how the workman feels toward his old employer.

Dependent upon the individual's situation, the service coordinator may make several visits to a workman's home over a period of time. However, a service coordinator never presents himself as Santa Claus.

From the very beginning it is made clear to a workman that the service cordinator is there to assist him in returning to gainful work. The workman is not promised a job nor any other sort of benefits he is not entitled to under the workmen's compensation law.

The service coordinator never discusses the workman's medical treatment. If a workman indicates he is unhappy with the medical care he is receiving, the service coordinator plainly advises the workman to discuss the problem with the treating physician.

One finds all the human problems present in working with the claimant workman. Sometimes a service coordinator can assist with the problem and sometimes not. For instance, a workman may be heavily indebted and the bill collectors are at his heels. This, plus his injury, may have the workman so depressed that he is not making any attempt to resume work.

A service coordinator may under these circumstances obtain an agreement with the credit agencies to stop their immediate demands and instead, formulate a plan wherein the workman can pay off his debts later when he is employed.

Through these contacts with the physician, the employer, and the workman, certain data are collected by the service coordinator. It is a serious error to create or demand that a service coordinator's time be consumed in completing a mound of paperwork and data collection. This is a common fallacy with agencies and it should be closely guarded against. The service coordinator's job is to spend time in behalf of the workman, not producing reports.

Useful Information

Oregon uses a one page standard reporting form which the service coordinator completes. The items common to most reporting forms of this nature that are found on this form are date referred, insurer, physician, diagnosis, date of injury, social security number, and such.

In addition, there are specific items which are quite useful to the service coordinator. Among these are:

Driver's license	Education
Own car	Technical training
Occupation and wage	Hobbies and interests
How long employed	Active union membership
Cause of injury	Military service disability
Other physical defects	Spouse's occupation and wage
Previous employers	Other income

There are specific reasons for gathering information on each of these items. All are used in assisting the service coordinator in appraising the situation and assisting the workman in a return to gainful occupation.

How does it help to know if the workman has a driver's license and a car? If the workman doesn't have these, there is no point in finding employment where the workman must commute unless there is good public transportation available. Nor can a workman be employed in a job that requires driving a vehicle unless a driver's license is obtained.

Looking at education and technical training, it is appalling how often this information is lacking in the claim files of workmen who are badly disabled over long periods of time. Physicians rarely ever report these facts and if they do, it is usually quite late in the claim. Insurers are equally guilty. If the intention is to prevent a workman's disability, it is imperative to know these facts at the start.

Other income and the spouse's income, if any, are important to know. Sometimes other income influences the workman's slowness to return to work. An example of this exists in the Oregon files in which the service coordinator overcame the problem.

In this instance the workman was a young man in his twenties. He was well-insured with disability insurance other than his workmen's compensation insurance. As a result, his income was over $500 a month and tax free.

In fact this workman's disability income more than equalled his take-home pay prior to his injury. It is no surprise that this young man had little inclination to return to work, even though his employer readily agreed to modify a job for him. Nor did his wife encourage him to resume work, according to the service coordinator, until one day.

On that day the service coordinator returned once again to the workman's home and had a serious talk with both the workman and his wife. The service coordinator clearly explained to this couple that, while they were living well without working at the time, one day it would all change rather suddenly.

The coordinator explained to them that the husband's treating physician would ultimately declare him medically stationary and able to work. When that happened, all their disability income would cease.

In addition, the service coordinator explained to the workman that his employer would not hold his job open forever, and if he didn't resume working, he would not have a job. With this the coordinator simply asked them, "What will you do when your insurance benefits are cut off and you don't have a job?"

This young workman and his wife reconsidered what they were doing and within a week he resumed working for his employer at a modified job. Had the service coordinator not known these facts, and had he not taken the action he did, this workman would have been rendered a disservice. It is very unlikely that any other party to the claim would have taken such action.

Military service disability can occasionally be valuable information for a service coordinator in the event vocational retraining is needed. An example is the situation in which the occupational injury and its residual effects are not sufficient to qualify the workman for vocational rehabilitation under workmen's compensation law.

Yet, if this workman continues his same work, it will in all probability lead to further injury and increasing disability. A service coordinator can guide this workman into vocational retraining under the Veteran's Administration benefits. In doing this the service coordinator is performing a function of both accident prevention and disability prevention.

Let's assume that a workman is eligible for vocational rehabilitation under workmen's compensation law. In this instance, there is no provision for maintenance provided by the workmen's compensation law for this workman during his vocational rehabilitation. Here again, Veteran's Administration benefits can be used to maintain this workman and his family until he completes his training.

Perhaps the workman is young and the best possible solution in returning him to gainful employment is through an apprenticeship program. If he is a young veteran of military service, this workman has a priority for admission in the apprenticeship program.

If a workman is an active union member, it is good to know

because union business agents and shop stewards can assist by requesting an employer to modify a job. Sometimes it is necessary to put the workman into a new job with the same employer. Here again, action by the union can make this possible.

RE-EMPLOYMENT

A decision which determines the course of events for a specific workman can be made immediately for the majority of claimants following the initial contacts with the physician, the employer, and the workman. For most, it is a matter of re-employment. For a few, it requires various levels of vocational retraining.

Looking at re-employment in an orderly and logical way, one can set down the choices available for most workmen:

1. Re-employment with same employer by job modification;
2. Re-employment with same employer in a new job without training;
3. Re-employment with same employer with on-the-job training;
4. Re-employment with a new employer in the same industry without training, or by job modification, or with on-the-job training;
5. Re-employment in a new industry without training;
6. Re-employment in a new industry by job modification, or with on-the-job training;
7. Re-employment by vocational training;
8. Re-employment by completion of formal education.

It is only sensible to attempt re-employment of a workman with the same employer providing there is a history of a good employee-employer relationship. The workman is naturally comfortable in this setting and he has fewer adjustments to make.

The next best approach is re-employment of a workman in the same industry although the employer is new. If employment in an industry new to a workman is necessary, the service coordinator's job becomes more critical. This brings us to a key factor in re-employment of workmen.

Good job placement is mandatory in any effective disability prevention program. It is not a matter of putting bodies to work on some kind of a job. It is the placement of a person, a workman, into a gainful work situation that is satisfactory to both the workman and the employer.

Poor job placement is a disservice to both a workman and an employer. In fact, it is dishonest because the workman and the employer both place their trust in the service coordinator when they commit themselves to the hire and the hiring. A workman wrongly placed in a job will not remain there, nor is the employer likely to hire another workman recommended by that service coordinator, or even the agency.

A knowledgeable service coordinator can avoid this problem most of the time. It is done by examining the workman's past, his jobs, his hobbies, his likes and dislikes. As an example, one does not take a workman such as a logger whose work life, interests, and hobbies are oriented to the out-of-doors, and put him to work as an orderly in a nursing home. This same workman who is a timber faller could be placed in millwork as a log scaler, or a lumber grader, and he will probably fit in very easily and be in his own element.

Service coordinators seek a job which fits both the nature of the workman and one which is within the limits of the workman's physical ability. It is always hoped that the job obtained for the workman is one with which the workman will remain.

Even short-term employment has merit. The psychological effect upon a workman who has a job is usually positive in nature. Should the workman choose to find a different job, the fact that he is already employed usually has a good psychological effect upon a potential employer.

An unemployed workman seeking a job usually has a strike against him when compared to a workman already working. An employer is likely to consider the workman who is employed as a better prospect where there is a choice.

On-the-job training (OJT) is sometimes required in placing a workman on a new job with the same employer or new employer. How the service coordinator accomplishes this is de-

pendent upon the laws of the state. A considerable amount of
responsibility is assumed by the service coordinator in making
a decision to place a workman in O.J.T. in Oregon.

Although all O.J.T. plans in Oregon are written by the Vocational Rehabilitation Division, the service coordinator makes
the actual arrangements. The service coordinator must be correct
in his judgment of the workman's ability to be trained for the
specific job.

If the service coordinator decides O.J.T. is feasible, he reaches
an agreement with the employer on the length of training required and the pay the workman is to receive.

On matters of a workman's wage during O.J.T., the service
coordinator is allowed to develop a subsidy agreement with the
employer, if the employer claims he cannot afford to pay a full
wage to the workman.

As an example of a wage agreement, let's assume a particular
workman is to receive O.J.T. for six months. The wage subsidy
agreement will probably be one in which the agency agrees to
reimburse the employer the following percentages of the wage
paid the workman:

1st month, 50% of the wage
2nd month, 35% of the wage
3rd month, 25% of the wage
4th month, 20% of the wage
5th month, 15% of the wage
6th month, 10% of the wage

Employers in Oregon are rarely reimbursed for more than 50
per cent of workman's wages during O.J.T. In addition, it is
rare for a workman to undergo O.J.T. for as long as one year.
The average O.J.T. plan is for three months.

If a service coordinator cannot make a clear-cut decision about
whether or not O.J.T. is suitable for a workman, he calls upon
other professionals to evaluate the workman for this purpose.

Vocational rehabilitation is probably not required as often as
many might believe. The present need to vocationally rehabilitate many workmen probably stems from the fact that we have

not utilized the concept of service coordinators. A service coordinator who functions effectively will reduce the need to vocationally rehabilitate as many workmen as we do today.

An underlying and perpetual problem present in our attempts to vocationally rehabilitate workmen is the tardiness of our efforts. Although a service coordinator is not directly involved in developing a vocational rehabilitation plan for a workman, the coordinator can cause it to be a timely effort.

There is absolutely nothing wrong in bringing the matter of vocational rehabilitation to the attention of the treating physician and the insurer. On the contrary, a service coordinator has this obligation if his evaluation of a workman causes the coordinator to believe this is the preferable choice.

At that time the service coordinator should revisit the treating physician and inquire if the physician concurs in the judgment. Because there is always a considerable time lag between the decision to vocationally rehabilitate and the implementation of a plan, such planning can begin prior to the workman's recovery of his ability to enter a plan.

It takes time to determine the suitability of a vocational rehabilitation plan. The retraining of a workman has to be in a field of work to his liking. For many workmen, this is not easy to determine because they are not certain of their own choices. The workman must be evaluated to determine whether or not he or she is capable of learning the job in mind.

There is little point in vocationally training a workman for a job in which the employment opportunity is nil. For this reason, service coordinators in Oregon use information on employment trends developed by the Division of Employment.

When talking with a workman about his possible need for vocational rehabilitation, the service coordinator tries to guide the workman's choices into a field with the best employment opportunities.

An important aspect of a service coordinator's role in vocational rehabilitation lies in determining the need early and persuading the parties involved to coordinate their actions on the workman's behalf for the workman's best interests.

JOB SEEKING, HOW AND WHERE

Jobs are where you find them. This is the philosophy of the Disability Prevention Division in Oregon. It is also another reason to hire workmen having broad work experiences as service coordinators.

Oregon's service coordinators have two general guidelines for job development:

1. Knock on doors;
2. Play it straight with the employer.

In such simple language, these suggestions do not appear as very high-minded guidelines. However, they work. Service coordinators utilize all sources for potential jobs for workmen. For example, the governor of Oregon pledged fifteen jobs for workmen to a service coordinator because of a chance meeting.

On this particular day, the governor and the service coordinator found themselves meeting in a hallway of the state capitol. They greeted one another and the governor, perhaps on impulse, began talking with the coordinator.

The service coordinator told the governor he would like an appointment to see him. He explained to the governor that he was developing a new program for the workmen's compensation agency and he believed the governor could help. The governor agreed to the meeting and the service coordinator made a thirty minute appointment to see the governor another day.

In his office, the governor became so interested in the service coordinator's explanation of the disability prevention program that the appointment lasted more than an hour and gained the needed job for some workmen.

A tendency exists whereby job developers look to large industries for jobs. While they are one job source, it is an error to consider them a major source of jobs. Oregon's service coordinators find that the major job source is small employers.

Playing it straight with an employer is a strict standard to adhere to in job placement. There are many reasons beyond the matter of simple honesty in doing this. Each employer hiring a workman through a service coordinator's efforts is also a potential future employer for another workman.

Bringing to an employer a workman well suited to the employer's needs is important for other reasons. Often an employer is asked to modify a job. This is an additional expense to an employer beyond the usual expenses of hiring a new employee.

Many employers are willing to take some gamble even though they are aware a problem exists. For this and other reasons, a service coordinator tells the potential employer all the facts he knows about a workman. The potential employer is made aware of all the known physical problems a workman may have, in addition to his work record.

The service coordinator's honesty with employers is paying dividends in Oregon. Employers are beginning to call Oregon's service coordinators as a means of finding a needed employee. A recent example of this occurred when an employer heard of the disability prevention program and it aroused his curiosity enough to cause the employer to visit the agency and inquire about the program. This employer needed an employee to do office work and bookkeeping.

After a lengthy discussion with a service coordinator, the employer asked the coordinator if there might be a workman available who could do bookkeeping and office work.

Four days prior to this employer's visit, the service coordinator visited a workman referred to the coordinator by an orthopedist. The workman was an amputee having a below-the-knee amputation of the right leg as a result of an industrial injury several years before.

The workman's present claim resulted from an injury to his left knee. For the past five years, he had worked on a dairy farm. Prior to this he was employed as a bookkeeper for a small business. When the owner sold the business, the new owner assumed the bookkeeping duties, putting the workman out of the job.

The employer who came to the agency in need of a bookkeeper owned a small but growing manufacturing and logging equipment repair plant. The service coordinator told this employer about the handicapped workman. An agreement for an on-the-job training plan was arranged with the employer.

Under this plan, the training period was set for six months. The beginning salary for this workman as a bookkeeper was $600 per month. The workman's previous wages on the dairy farm were $400 per month. The agency paid 50 per cent of the workman's wages the first month. After this, the agency wage contribution declined proportionately each month until the seventh month when the employer began paying the workman's full wage.

Good job placement is also accident prevention. This is why a potential employer is made fully aware of a workman's physical condition. An example of this exists in the claim file of a 31-year old auto mechanic.

This workman injured his back and left leg one day while replacing an automobile engine. His treating orthopedist found that this man had a spondylolisthesis. In the orthopedist's opinion, if the workman continued his present job, he would ultimately be disabled by his back condition.

The orthopedist recommended a change in jobs for the workman. For some reason, the insurer overlooked this recommendation. Three weeks later, the workman asked the insurer for help in vocational training.

The insurer ignored the workman's request and instead referred this man to the Workmen's Compensation Board's physical rehabilitation center for evaluation, presumably hoping to close the claim. At this point five months after his injury, the workman was referred by the physical rehabilitation center personnel to a service coordinator.

The service coordinator began working with the workman, the orthopedist, and the workman's employer. The employer had no jobs which the orthopedist believed suitable for the workman.

A new job requiring on-the-job training as an auto parts counterman was found for the workman. Since this workman was already a mechanic, the new employer and the service coordinator concluded that three months training would suffice.

On followup, this workman is fully employed making a decent living in a job he likes. The job is also within the limits of his physical capacity. The new employer is pleased with the workman and considers him a good employee. In addition, this workman is working at a job within his physical limitations. He is less apt to reinjure his back and become one of the repeated back injury claimants seen so often in workmen's compensation.

It is impossible to legislate workmen's compensation laws that prevent disability. Nor is it possible to prevent disability by merely establishing a disability prevention division. It takes people, intelligent men and women having both an interest in working with people and plenty of experience in the work world to get the job of disability prevention done.

So long as people earn by occupation, a need to prevent disability exists for those made ill or injured in their jobs. Most states' laws already provide those things that help to prevent disability. Whether or not these provisions are properly delivered determines whether a workman is disabled or not in fact.

The workmen's compensation agencies of the states are capable of taking the initiative in disability prevention. It takes more than that though for disability prevention to be a reality in a state. It requires the support of all people in a state, leaders and citizens alike. It is they who determine the outcome of an injured workman in their state, which is another way of saying, "It is up to you and me."

BIBLIOGRAPHY

1. Chamber of Commerce of the United States: *Analysis of Workmen's Compensation Laws.* Washington, D.C., 1972 and 1973 ed.
2. Report of the National Commission on State Workmen's Compensation Laws: The report of. Washington, D.C., 1972.
3. National Safety Council: *Accident Facts.* Chicago, 1972 ed.
4. McCallister, R. L., Manager Workmen's Compensation Claims, Georgia Pacific Corporation. Testimony before National Commission, February 21, 1972.
5. Skelton, Keith D., Professor of Law at Portland State University, Representative Oregon Legislature. Testimony before National Commission, November 15, 1971.
6. Workmen's Compensation Bureau, Florida Dept. of Commerce: *Cases, Causes, Costs, 1972.*
7. Workmen's Compensation Board of the State of Oregon: *Timeliness of First Payments.* Salem, Oregon, 1972.
8. Page, Joseph A. and O'Brien, Mary Win: Statement on Occupational Safety and Health before The National Commission on State Workmen's Compensation Laws. Washington, D.C., September 23, 1971.
9. Leavitt, Stephen S., *et al.*: The process of recovery: Patterns in industrial back injury. *Industrial Medicine, Part I, 40*(8):7, 1971; Part II, *40*(9):7, 1971; Part III, *41*(1):7, 1972; Part IV, *41*(2):5, 1972.
10. Webb, Jack: *Permanent Partial Disability, Past Proposals and Future Hopes.* Panel discussion at annual convention of International Association of Industrial Accident Boards and Commissions, Boston, 1971.
11. Beals, Rodney K. and Hickman, Norman W.: Industrial injuries of the back and extremities. *The Journal of Bone and Joint Surgery, 54-A* (8):1593, 1972.
12. Meineker, Robert L.: Psychology of Work. *Journal of Occupational Medicine, 14*(3):212, 1972.
13. Fisher, Dennis U.: Preliminary Report for the Research Project: An Analysis of Rates and Provisions of Workmen's Compensation Insurance Carried by Farmers in Oregon. Salem, Oregon, January 29, 1973.

14. Schramm, Carl J. (with the assistance of Marion Pitts): *Workmen's Compensation: The Identification of Potentital Rehabilitants.* Washington, D.C.: The National Commission on State Workmen's Compensation Laws, Document #64.

15. Martin, R. A.: Permanent Total Disability from Low Back Injury. A study for the Workmen's Compensation Board, Oregon, 1969.

16. Wise, Paul S.: *The Challenge of Workmen's Compensation: Six Guidelines for Action.* Speech to the International Association of Industrial Accident Boards and Commissions, Boston, September 13, 1971.

17. White, A. W. M.: Low back pain in men receiving workmen's compensation. *Canadian Medical Association Journal, 95*: 50, 1966.

18. White, A. W. M.: Low back pain in men receiving workmen's compensation. A follow-up study. *Canadian Medical Association Journal, 101*:61, 1969.